Habitats Local and Far Away, Grade 1

What if you could challenge your first graders to imagine saving an endangered species, learning about different global habitats along the way? With this volume in the *STEM Road Map Curriculum Series*, you can! *Habitats Local and Far Away* outlines a journey that will steer your students toward authentic problem solving while grounding them in integrated STEM disciplines. Like the other volumes in the series, this book is designed to meet the growing need to infuse real-world learning into K–12 classrooms.

This interdisciplinary, four-lesson module uses project- and problem-based learning to help students develop an action plan to encourage preservation of an endangered species. Students will work in teams to describe the habitat characteristics of a species outside their home region, explain why the species is endangered, and offer solutions about how humans might be able to support this species' survival. In developing their plan, they will act as explorers of species locally and around the world, learning about climate, plant and animal inhabitants, and key factors affecting habitat vitality or decline.

To support this goal, students will do the following:

- Explain that there are various types of habitats that vary with geographical location around the world

- Identify several habitats in the U.S. and globally

- Explain how various habitats meet animals' basic needs

- Identify climatic characteristics of several habitats

- Identify humans as species that live within and in interaction with various habitats

- Identify technological advances and tools that scientists use to learn about habitats and endangered species

- Design and construct models to demonstrate understanding of features of various habitats (local and global) and endangered species

- Apply their knowledge of habitat characteristics, interdependence in ecosystems, and endangered species to develop an action plan to help preserve their selected endangered species

The *STEM Road Map Curriculum Series* is anchored in the Next Generation Science Standards, the Common Core State Standards, and the Framework for 21st Century Learning. In-depth and flexible, *Habitats Local and Far Away* can be used as a whole unit or in part to meet the needs of districts, schools, and teachers who are charting a course toward an integrated STEM approach.

Carla C. Johnson is a Professor of Science Education and Office of Research and Innovation Faculty Research Fellow at North Carolina State University, North Carolina, USA.

Janet B. Walton is a Senior Research Scholar at North Carolina State's College of Education in Raleigh, North Carolina, USA.

Erin E. Peters-Burton is the Donna R. and David E. Sterling Endowed Professor in Science Education at George Mason University in Fairfax, Virginia, USA.

THE STEM ROAD MAP CURRICULUM SERIES

Series editors: Carla C. Johnson, Janet B. Walton, and Erin E. Peters-Burton

Map out a journey that will steer your students toward authentic problem solving as you ground them in integrated STEM disciplines.

Co-published by Routledge and NSTA Press, in partnership with the National Science Teaching Association, this K–12 curriculum series is anchored in the Next Generation Science Standards, the Common Core State Standards, and the Framework for 21st Century Learning. It was developed to meet the growing need to infuse real-world STEM learning into classrooms.

Each book is an in-depth module that uses project- and problem-based learning. First, your students are presented with a challenge. Then, they apply what they learn using science, social studies, English language arts, and mathematics. Engaging and flexible, each volume can be used as a whole unit or in part to meet the needs of districts, schools, and teachers who are charting a course toward an integrated STEM approach.

Modules are available from NSTA Press and Routledge, and organized under the following themes. For an update listing of the volumes in the series, please visit https://www.routledge.com/STEM-Road-Map-Curriculum-Series/book-series/SRM (for titles co-published by Routledge and NSTA Press), or www.nsta.org/book-series/stem-road-map-curriculum (for titles published by NSTA Press).

Co-published by Routledge and NSTA Press:

Optimizing the Human Experience:

- *Our Changing Environment, Grade K: STEM Road Map for Elementary School*
- *Genetically Modified Organisms, Grade 7: STEM Road Map for Middle School*
- *Rebuilding the Natural Environment, Grade 10: STEM Road Map for High School*
- *Mineral Resources, Grade 11: STEM Road Map for High School*

Cause and Effect:

- *Formation of the Earth, Grade 9: STEM Road Map for High School*

Sustainable Systems:

- *Habitats in the United States, Grade K: STEM Road Map for Elementary School*
- *Habitats Local and Far Away, Grade 1: STEM Road Map for Elementary School*
- *Hydropower Efficiency, Grade 4: STEM Road Map for Elementary School*
- *Composting, Grade 5: STEM Road Map for Elementary School*
- *Global Population Issues, Grade 7: STEM Road Map for Middle School*
- *The Speed of Green, Grade 8: STEM Road Map for Middle School*
- *Creating Global Bonds, Grade 12: STEM*
- *Road Map for High School*

Published by NSTA Press:

Innovation and Progress:

- *Amusement Park of the Future, Grade 6: STEM Road Map for Elementary School*
- *Transportation in the Future, Grade 3: STEM Road Map for Elementary School*

- *Harnessing Solar Energy, Grade 4: STEM Road Map for Elementary School*
- *Wind Energy, Grade 5: STEM Road Map for Elementary School*
- *Construction Materials, Grade 11: STEM Road Map for High School*

The Represented World:

- *Patterns and the Plant World, Grade 1: STEM Road Map for Elementary School*
- *Investigating Environmental Changes, Grade 2: STEM Road Map for Elementary School*
- *Swing Set Makeover, Grade 3: STEM Road Map for Elementary School*
- *Rainwater Analysis, Grade 5: STEM Road Map for Elementary School*
- *Packaging Design, Grade 6: STEM Road Map for Middle School*
- *Improving Bridge Design, Grade 8: STEM Road Map for Middle School*
- *Radioactivity, Grade 11: STEM Road Map for High School*
- *Car Crashes, Grade 12: STEM Road Map for High School*

Cause and Effect:

- *Physics in Motion, Grade K: STEM Road Map for Elementary School*
- *Influence of Waves, Grade 1: STEM Road Map for Elementary School*
- *Natural Hazards, Grade 2: STEM Road Map for Elementary School*
- *Human Impacts on Our Climate, Grade 6: STEM Road Map for Middle School*
- *The Changing Earth, Grade 8: STEM Road Map for Middle School*
- *Healthy Living, Grade 10: STEM Road Map for High School*

Habitats Local and Far Away, Grade 1

Grade 1

STEM Road Map for Elementary School

Edited by Carla C. Johnson, Janet B. Walton, and Erin E. Peters-Burton

Routledge
Taylor & Francis Group

NEW YORK AND LONDON

nsta Press
National Science Teaching Association

Designed cover image: © Shutterstock and © Getty Images

First published 2024
by Routledge
605 Third Avenue, New York, NY 10158

and by Routledge
4 Park Square, Milton Park, Abingdon, Oxon, OX14 4RN

Routledge is an imprint of the Taylor & Francis Group, an informa business

A co-publication with NSTA Press

Routledge is committed to publishing material that promotes the best in inquiry-based science education. However, conditions of actual use may vary, and the safety procedures and practices described in this book are intended to serve only as a guide. Additional precautionary measures may be required. Routledge and the authors do not warrant or represent that the procedures and practices in this book meet any safety code or standard of federal, state, or local regulations. Routledge and the authors disclaim any liability for personal injury or damage to property arising out of or relating to the use of this book, including any of the recommendations, instructions, or materials contained therein.

Trademark notice: Product or corporate names may be trademarks or registered trademarks, and are used only for identification and explanation without intent to infringe.

Library of Congress Cataloging-in-Publication Data
Names: Johnson, Carla C., 1969– editor. | Walton, Janet B., 1968– editor. | Peters-Burton, Erin E., editor.
Title: Habitats local and far away, grade 1 : STEM road map for elementary school / edited by Carla C. Johnson, Janet B. Walton, and Erin E. Peters-Burton.
Description: New York, NY : Routledge, 2024. | Series: STEM road map curriculum series | Includes bibliographical references and index.
Identifiers: LCCN 2023042178 (print) | LCCN 2023042179 (ebook) | ISBN 9781032584645 (hardback) | ISBN 9781032584676 (paperback) | ISBN 9781003450184 (ebook)
Subjects: LCSH: Habitat (Ecology)—Study and teaching (Elementary) | First grade (Education) | Habitat (Ecology)—Study and teaching (Elementary)—Activity programs | Habitat (Ecology)—Study and teaching (Primary) | Ecology—Study and teaching (Elementary)—Activity programs.
Classification: LCC QH541.2 .H336 2024 (print) | LCC QH541.2 (ebook) | DDC 577.071—dc23/eng/20231222
LC record available at https://lccn.loc.gov/2023042178
LC ebook record available at https://lccn.loc.gov/2023042179

ISBN: 978-1-032-584645 (hbk)
ISBN: 978-1-032-584676 (pbk)
ISBN: 978-1-003-450184 (ebk)

DOI: 10.4324/9781003450184

Typeset in Palatino
by Apex CoVantage, LLC

CONTENTS

Part 1: The STEM Road Map: Background, Theory, and Practice

Part 2: Habitats Local and Far Away: STEM Road Map Module

CONTENTS

4 Habitats Local and Far Away Lesson Plans 39

*Andrea R. Milner, Vanessa B. Morrison, Janet B. Walton,
Carla C. Johnson, and Erin E. Peters-Burton*

CONTENTS

ABOUT THE EDITORS AND AUTHORS

Dr. Carla C. Johnson is a professor of science education and Office of Research and Innovation Faculty Research Fellow at NC State University. Dr. Johnson has served (2015–2021) as the director of research and evaluation for the Department of Defense–funded Army Educational Outreach Program (AEOP), a global portfolio of STEM education programs, competitions, and apprenticeships. She has been a leader in STEM education for the past decade, serving as the director of STEM Centers, editor of the *School Science and Mathematics* journal, and lead researcher for the evaluation of Tennessee's Race to the Top–funded STEM portfolio. Dr. Johnson has published over 200 articles, books, book chapters, and curriculum books focused on STEM education. She is a former science and social studies teacher and was the recipient of the 2013 Outstanding Science Teacher Educator of the Year award from the Association for Science Teacher Education (ASTE), the 2012 Award for Excellence in Integrating Science and Mathematics from the School Science and Mathematics Association (SSMA), the 2014 award for best paper on Implications of Research for Educational Practice from ASTE, and the 2006 Outstanding Early Career Scholar Award from SSMA. Her research focuses on STEM education policy implementation, effective science teaching, and integrated STEM approaches.

Dr. Janet B. Walton is a senior research scholar at NC State University's College of Education in Raleigh, North Carolina. Dr. Walton served as assistant director of research and evaluation for the Army Educational Outreach Program (AEOP) from 2015 through 2021. She leverages backgrounds in economic development and education to develop K–12 curricular materials that integrate real-life issues with sound cross-curricular content and provide students and educators with innovative resources and curricular materials. Her research focuses includes collaboration between schools and community stakeholders for STEM education, problem- and project- based learning pedagogies, online learning, and mixed methods research methodologies.

Dr. Erin E. Peters-Burton is the Donna R. and David E. Sterling endowed professor in science education at George Mason University in Fairfax, Virginia. She uses her experiences from 15 years as an engineer and secondary science, engineering, and mathematics teacher to develop research projects that directly inform classroom practice in science and engineering. Her research agenda is based on the idea that all students

should build self-awareness of how they learn science and engineering. She works to help students see themselves as "science-minded" and help teachers create classrooms that support student skills to develop scientific knowledge. To accomplish this, she pursues research projects that investigate ways that students and teachers can use self-regulated learning theory in science and engineering, as well as how inclusive STEM schools can help students succeed. She received the Outstanding Science Teacher Educator of the Year award from ASTE in 2016 and a Teacher of Distinction Award and a Scholarly Achievement Award from George Mason University in 2012, and in 2010 she was named University Science Educator of the Year by the Virginia Association of Science Teachers.

Dr. Toni A. May is an associate professor of assessment, research, and statistics in the School of Education at Drexel University in Philadelphia. Dr. May's research concentrates on assessment and evaluation in education, with a focus on K–12 STEM.

Dr. Andrea R. Milner is the vice president and dean of academic affairs and an associate professor in the Teacher Education Department at Adrian College in Adrian, Michigan. A former early childhood and elementary teacher, Dr. Milner researches the effects constructivist classroom contextual factors have on student motivation and learning strategy use.

Dr. Tamara J. Moore is an associate professor of engineering education in the College of Engineering at Purdue University. Dr. Moore's research focuses on defining STEM integration through the use of engineering as the connection and investigating its power for student learning.

Dr. Vanessa B. Morrison is an associate professor in the Teacher Education Department at Adrian College. She is a former early childhood teacher and reading and language arts specialist whose research is focused on learning and teaching within a transdisciplinary framework.

ACKNOWLEDGMENTS

This module was developed as a part of the STEM Road Map project (Carla C. Johnson, principal investigator). The Purdue University College of Education, General Motors, and other sources provided funding for this project.

See www.routledge.com/9780367467524 for more information about *STEM Road Map: A Framework for Integrated STEM Education.*

PART 1

THE STEM ROAD MAP

BACKGROUND, THEORY, AND PRACTICE

OVERVIEW OF THE *STEM ROAD MAP CURRICULUM SERIES*

Carla C. Johnson, Erin E. Peters-Burton, and Tamara J. Moore

The *STEM Road Map Curriculum Series* was conceptualized and developed by a team of STEM educators from across the United States in response to a growing need to infuse real-world learning contexts, delivered through authentic problem-solving pedagogy, into K–12 classrooms. The curriculum series is grounded in integrated STEM, which focuses on the integration of the STEM disciplines –science, technology, engineering, and mathematics – delivered across content areas, incorporating the Framework for 21st Century Learning along with grade-level-appropriate academic standards. The curriculum series begins in kindergarten, with a five-week instructional sequence that introduces students to the STEM themes and gives them grade-level-appropriate topics and real-world challenges or problems to solve. The series uses project-based and problem-based learning, presenting students with the problem or challenge during the first lesson, and then teaching them science, social studies, English language arts, mathematics, and other content, as they apply what they learn to the challenge or problem at hand.

Authentic assessment and differentiation are embedded throughout the modules. Each *STEM Road Map Curriculum Series* module has a lead discipline, which may be science, social studies, English language arts, or mathematics. All disciplines are integrated into each module, along with ties to engineering. Another key component is the use of STEM Research Notebooks to allow students to track their own learning progress. The modules are designed with a scaffolded approach, with increasingly complex concepts and skills introduced as students progress through grade levels.

The developers of this work view the curriculum as a resource that is intended to be used either as a whole or in part to meet the needs of districts, schools, and teachers who are implementing an integrated STEM approach. A variety of implementation formats are possible, from using one stand-alone module at a given grade level to using all five modules to provide 25 weeks of instruction. Also, within each grade band (K–2, 3–5, 6–8, 9–12), the modules can be sequenced in various ways to suit specific needs.

DOI: 10.4324/9781003450184-2

STANDARDS-BASED APPROACH

The *STEM Road Map Curriculum Series* is anchored in the *Next Generation Science Standards (NGSS)*, the *Common Core State Standards for Mathematics (CCSS Mathematics)*, the *Common Core State Standards for English Language Arts (CCSS ELA)*, and the Framework for 21st Century Learning. Each module includes a detailed curriculum map that incorporates the associated standards from the particular area correlated to lesson plans. The STEM Road Map has very clear and strong connections to these academic standards, and each of the grade-level topics was derived from the mapping of the standards to ensure alignment among topics, challenges or problems, and the required academic standards for students. Therefore, the curriculum series takes a standards-based approach and is designed to provide authentic contexts for application of required knowledge and skills.

THEMES IN THE *STEM ROAD MAP CURRICULUM SERIES*

The K–12 STEM Road Map is organized around five real-world STEM themes that were generated through an examination of the big ideas and challenges for society included in STEM standards and those that are persistent dilemmas for current and future generations:

- Cause and Effect
- Innovation and Progress
- The Represented World
- Sustainable Systems
- Optimizing the Human Experience

These themes are designed as springboards for launching students into an exploration of real-world learning situated within big ideas. Most important, the five STEM Road Map themes serve as a framework for scaffolding STEM learning across the K–12 continuum.

The themes are distributed across the STEM disciplines so that they represent the big ideas in science (Cause and Effect; Sustainable Systems), technology (Innovation and Progress; Optimizing the Human Experience), engineering (Innovation and Progress; Sustainable Systems; Optimizing the Human Experience), and mathematics (The Represented World), as well as concepts and challenges in social studies and 21st century skills that are also excellent contexts for learning in English language arts. The process of developing themes began with the clustering of the *NGSS* performance expectations and the National Academy of Engineering's grand challenges for engineering, which led to the development of the challenge in each module and connections of the module activities to the *CCSS Mathematics* and *CCSS ELA* standards. We performed

these mapping processes with large teams of experts and found that these five themes provided breadth, depth, and coherence to frame a high-quality STEM learning experience from kindergarten through 12th grade.

Cause and Effect

The concept of cause and effect is a powerful and pervasive notion in the STEM fields. It is the foundation of understanding how and why things happen as they do. Humans spend considerable effort and resources trying to understand the causes and effects of natural and designed phenomena to gain better control over events and the environment and to be prepared to react appropriately. Equipped with the knowledge of a specific cause-and-effect relationship, we can lead better lives or contribute to the community by altering the cause, leading to a different effect. For example, if a person recognizes that irresponsible energy consumption leads to global climate change, that person can act to remedy his or her contribution to the situation. Although cause and effect is a core idea in the STEM fields, it can actually be difficult to determine. Students should be capable of understanding not only when evidence points to cause and effect but also when evidence points to relationships but not direct causality. The major goal of education is to foster students to be empowered, analytic thinkers, capable of thinking through complex processes to make important decisions. Understanding causality, as well as when it cannot be determined, will help students become better consumers, global citizens, and community members.

Innovation and Progress

One of the most important factors in determining whether humans will have a positive future is innovation. Innovation is the driving force behind progress, which helps create possibilities that did not exist before. Innovation and progress are creative entities, but in the STEM fields they are anchored by evidence and logic, and they use established concepts to move the STEM fields forward. In creating something new, students must consider what is already known in the STEM fields and apply this knowledge appropriately. When we innovate, we create value that was not there previously and create new conditions and possibilities for even more innovations. Students should consider how their innovations might affect progress and use their STEM thinking to change current human burdens to benefits. For example, if we develop more efficient cars that use by-products from another manufacturing industry, such as food processing, then we have used waste productively and reduced the need for the waste to be hauled away, an indirect benefit of the innovation.

The Represented World

When we communicate about the world we live in, how the world works, and how we can meet the needs of humans, sometimes we can use the actual phenomena to

explain a concept. Sometimes, however, the concept is too big, too slow, too small, too fast, or too complex for us to explain using the actual phenomena, and we must use a representation or a model to help communicate the important features. We need representations and models such as graphs, tables, mathematical expressions, and diagrams because it makes our thinking visible. For example, when examining geologic time, we cannot actually observe the passage of such large chunks of time, so we create a timeline or a model that uses a proportional scale to visually illustrate how much time has passed for different eras. Another example may be something too complex for students at a particular grade level, such as explaining the p subshell orbitals of electrons to fifth graders. Instead, we use the Bohr model, which more closely represents the orbiting of planets and is accessible to fifth graders.

When we create models, they are helpful because they point out the most important features of a phenomenon. We also create representations of the world with mathematical functions, which help us change parameters to suit the situation. Creating representations of a phenomenon engages students because they are able to identify the important features of that phenomenon and communicate them directly. But because models are estimates of a phenomenon, they leave out some of the details, so it is important for students to evaluate their usefulness as well as their shortcomings.

Sustainable Systems

From an engineering perspective, the term *system* refers to the use of "concepts of component need, component interaction, systems interaction, and feedback. The interaction of subcomponents to produce a functional system is a common lens used by all engineering disciplines for understanding, analysis, and design" (Koehler, Bloom, and Binns 2013, p. 8). Systems can be either open (e.g., an ecosystem) or closed (e.g., a car battery). Ideally, a system should be sustainable, able to maintain equilibrium without much energy from outside the structure. Looking at a garden, we see flowers blooming, weeds sprouting, insects buzzing, and various forms of life living within its boundaries. This is an example of an ecosystem, a collection of living organisms that survive together, functioning as a system. The interaction of the organisms within the system and the influences of the environment (e.g., water, sunlight) can maintain the system for a period of time, thus demonstrating its ability to endure. Sustainability is a desirable feature of a system because it allows for existence of the entity in the long term.

In the STEM Road Map project, we identified different standards that we consider to be oriented toward systems that students should know and understand in the K–12 setting. These include ecosystems, the rock cycle, Earth processes (such as erosion, tectonics, ocean currents, weather phenomena), Earth-Sun-Moon cycles, heat transfer, and the interaction among the geosphere, biosphere, hydrosphere, and atmosphere. Students and teachers should understand that we live in a world of systems that are not independent of each other, but rather are intrinsically linked such that a disruption in one part of a system will have reverberating effects on other parts of the system.

Optimizing the Human Experience

Science, technology, engineering, and mathematics as disciplines have the capacity to continuously improve the ways humans live, interact, and find meaning in the world, thus working to optimize the human experience. This idea has two components: being more suited to our environment and being more fully human. For example, the progression of STEM ideas can help humans create solutions to complex problems, such as improving ways to access water sources, designing energy sources with minimal impact on our environment, developing new ways of communication and expression, and building efficient shelters. STEM ideas can also provide access to the secrets and wonders of nature. Learning in STEM requires students to think logically and systematically, which is a way of knowing the world that is markedly different from knowing the world as an artist. When students can employ various ways of knowing and understand when it is appropriate to use a different way of knowing or integrate ways of knowing, they are fully experiencing the best of what it is to be human. The problem-based learning scenarios provided in the STEM Road Map help students develop ways of thinking like STEM professionals as they ask questions and design solutions. They learn to optimize the human experience by innovating improvements in the designed world in which they live.

THE NEED FOR AN INTEGRATED STEM APPROACH

At a basic level, STEM stands for science, technology, engineering, and mathematics. Over the past decade, however, STEM has evolved to have a much broader scope and implications. Now, educators and policy makers refer to STEM as not only a concentrated area for investing in the future of the United States and other nations but also as a domain and mechanism for educational reform.

The good intentions of the recent decade-plus of focus on accountability and increased testing has resulted in significant decreases not only in instructional time for teaching science and social studies but also in the flexibility of teachers to promote authentic, problem-solving–focused classroom environments. The shift has had a detrimental impact on student acquisition of vitally important skills, which many refer to as 21st century skills, and often the ability of students to "think." Further, schooling has become increasingly siloed into compartments of mathematics, science, English language arts, and social studies, lacking any of the connections that are overwhelmingly present in the real world around children. Students have experienced school as content provided in boxes that must be memorized, devoid of any real-world context, and often have little understanding of why they are learning these things.

STEM-focused projects, curriculum, activities, and schools have emerged as a means to address these challenges. However, most of these efforts have continued to focus on the individual STEM disciplines (predominantly science and engineering) through more STEM classes and after-school programs in a "STEM enhanced" approach (Breiner et al., 2012). But in traditional and STEM-enhanced approaches, there is little to no focus on other disciplines that are integral to the context of STEM in the real world. Integrated STEM education, on the other hand, infuses the learning of important STEM content and concepts with a much-needed emphasis on 21st century skills and a problem- and project-based pedagogy that more closely mirrors the real-world setting for society's challenges. It incorporates social studies, English language arts, and the arts as pivotal and necessary (Johnson 2013; Rennie, Venville, and Wallace 2012; Roehrig et al. 2012).

FRAMEWORK FOR STEM INTEGRATION IN THE CLASSROOM

The *STEM Road Map Curriculum Series* is grounded in the Framework for STEM Integration in the Classroom as conceptualized by Moore, Guzey, and Brown (2014) and Moore et al. (2014). The framework has six elements, described in the context of how they are used in the *STEM Road Map Curriculum Series* as follows:

1. The STEM Road Map contexts are meaningful to students and provide motivation to engage with the content. Together, these allow students to have different ways to enter into the challenge.

2. The STEM Road Map modules include engineering design that allows students to design technologies (i.e., products that are part of the designed world) for a compelling purpose.

3. The STEM Road Map modules provide students with the opportunities to learn from failure and redesign based on the lessons learned.

4. The STEM Road Map modules include standards-based disciplinary content as the learning objectives.

5. The STEM Road Map modules include student-centered pedagogies that allow students to grapple with the content, tie their ideas to the context, and learn to think for themselves as they deepen their conceptual knowledge.

6. The STEM Road Map modules emphasize 21st century skills and, in particular, highlight communication and teamwork.

All of the STEM Road Map modules incorporate these six elements; however, the level of emphasis on each of these elements varies based on the challenge or problem in each module.

THE NEED FOR THE *STEM ROAD MAP CURRICULUM SERIES*

As focus is increasing on integrated STEM, and additional schools and programs decide to move their curriculum and instruction in this direction, there is a need for high-quality, research-based curriculum designed with integrated STEM at the core. Several good resources are available to help teachers infuse engineering or more STEM enhanced approaches, but no curriculum exists that spans K–12 with an integrated STEM focus. The next chapter provides detailed information about the specific pedagogy, instructional strategies, and learning theory on which the *STEM Road Map Curriculum Series* is grounded.

REFERENCES

Breiner, J., M. Harkness, C. C. Johnson, and C. Koehler. 2012. What is STEM? A discussion about conceptions of STEM in education and partnerships. *School Science and Mathematics* 112 (1): 3–11.

Johnson, C. C. 2013. Conceptualizing integrated STEM education: Editorial. *School Science and Mathematics* 113 (8): 367–368.

Koehler, C. M., M. A. Bloom, and I. C. Binns. 2013. Lights, camera, action: Developing a methodology to document mainstream films' portrayal of nature of science and scientific inquiry. *Electronic Journal of Science Education* 17 (2).

Moore, T. J., S. S. Guzey, and A. Brown. 2014. Greenhouse design to increase habitable land: An engineering unit. *Science Scope* 37 (7): 51–57.

Moore, T. J., M. S. Stohlmann, H.-H. Wang, K. M. Tank, A. W. Glancy, and G. H. Roehrig. 2014. Implementation and integration of engineering in K–12 STEM education. In *Engineering in pre-college settings: Synthesizing research, policy, and practices*, ed. S. Purzer, J. Strobel, and M. Cardella, 35–60. West Lafayette, IN: Purdue Press.

Rennie, L., G. Venville, and J. Wallace. 2012. *Integrating science, technology, engineering, and mathematics: Issues, reflections, and ways forward.* New York: Routledge.

Roehrig, G. H., T. J. Moore, H. H. Wang, and M. S. Park. 2012. Is adding the *E* enough? Investigating the impact of K–12 engineering standards on the implementation of STEM integration. *School Science and Mathematics* 112 (1): 31–44.

STRATEGIES USED IN *THE STEM ROAD MAP CURRICULUM SERIES*

Erin E. Peters-Burton, Carla C. Johnson, Toni A. Sondergeld, and Tamara J. Moore

The *STEM Road Map Curriculum Series* uses what has been identified through research as best-practice pedagogy, including embedded formative assessment strategies throughout each module. This chapter briefly describes the key strategies that are employed in the series.

PROJECT- AND PROBLEM-BASED LEARNING

Each module in the *STEM Road Map Curriculum Series* uses either project-based learning or problem-based learning to drive the instruction. Project-based learning begins with a driving question to guide student teams in addressing a contextualized local or community problem or issue. The outcome of project-based instruction is a product that is conceptualized, designed, and tested through a series of scaffolded learning experiences (Blumenfeld et al. 1991; Krajcik and Blumenfeld 2006). Problem-based learning is often grounded in a fictitious scenario, challenge, or problem (Barell 2006; Lambros 2004). On the first day of instruction within the unit, student teams are provided with the context of the problem. Teams work through a series of activities and use open-ended research to develop their potential solution to the problem or challenge, which need not be a tangible product (Johnson, 2003).

ENGINEERING DESIGN PROCESS

The *STEM Road Map Curriculum Series* uses engineering design as a way to facilitate integrated STEM within the modules. The engineering design process (EDP) is depicted in Figure 2.1 (p. 10). It highlights two major aspects of engineering design – problem scoping and solution generation – and six specific components of working

Habitats Local and Far Away, Grade 1

9

DOI: 10.4324/9781003450184-3

Figure 2.1. Engineering Design Process

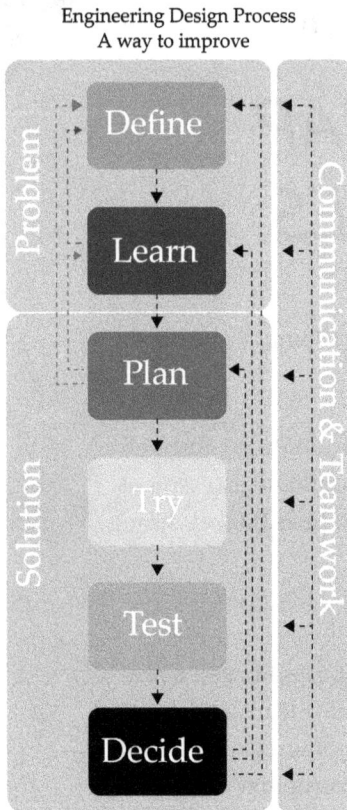

Engineering Design Process
A way to improve

Define

Learn

Plan

Try

Test

Decide

Problem

Solution

Communication & Teamwork

toward a design: define the problem, learn about the problem, plan a solution, try the solution, test the solution, decide whether the solution is good enough. It also shows that communication and teamwork are involved throughout the entire process. As the arrows in the figure indicate, the order in which the components of engineering design are addressed depends on what becomes needed as designers progress through the EDP. Designers must communicate and work in teams throughout the process. The EDP is iterative, meaning that components of the process can be repeated as needed until the design is good enough to present to the client as a potential solution to the problem.

Problem scoping is the process of gathering and analyzing information to deeply understand the engineering design problem. It includes defining the problem and learning about the problem. Defining the problem includes identifying the problem, the client, and the end user of the design. The client is the person (or people) who hired the designers to do the work, and the end user is the person (or people) who will use the final design. The designers must also identify the criteria and the constraints of the problem. The criteria are the things the client wants from the solution, and the constraints are the things that limit the possible solutions. The designers must spend significant time learning about the problem, which can include activities such as the following:

- Reading informational texts and researching about relevant concepts or contexts

- Identifying and learning about needed mathematical and scientific skills, knowledge, and tools

- Learning about things done previously to solve similar problems

- Experimenting with possible materials that could be used in the design

Problem scoping also allows designers to consider how to measure the success of the design in addressing specific criteria and staying within the constraints over multiple iterations of solution generation.

Solution generation includes planning a solution, trying the solution, testing the solution, and deciding whether the solution is good enough. Planning the solution includes generating many design ideas that both address the criteria and meet the constraints.

Here the designers must consider what was learned about the problem during problem scoping. Design plans include clear communication of design ideas through media such as notebooks, blueprints, schematics, or storyboards. They also include details about the design, such as measurements, materials, colors, costs of materials, instructions for how things fit together, and sets of directions. Making the decision about which design idea to move forward involves considering the trade-offs of each design idea.

Once a clear design plan is in place, the designers must try the solution. Trying the solution includes developing a prototype (a testable model) based on the plan generated. The prototype might be something physical or a process to accomplish a goal. This component of design requires that the designers consider the risk involved in implementing the design. The prototype developed must be tested. Testing the solution includes conducting fair tests that verify whether the plan is a solution that is good enough to meet the client and end user needs and wants. Data need to be collected about the results of the tests of the prototype, and these data should be used to make evidence-based decisions regarding the design choices made in the plan. Here, the designers must again consider the criteria and constraints for the problem.

Using the data gathered from the testing, the designers must decide whether the solution is good enough to meet the client and end user needs and wants by assessment based on the criteria and constraints. Here, the designers must justify or reject design decisions based on the background research gathered while learning about the problem and on the evidence gathered during the testing of the solution. The designers must now decide whether to present the current solution to the client as a possibility or to do more iterations of design on the solution. If they decide that improvements need to be made to the solution, the designers must decide if there is more that needs to be understood about the problem, client, or end user; if another design idea should be tried; or if more planning needs to be conducted on the same design. One way or another, more work needs to be done.

Throughout the process of designing a solution to meet a client's needs and wants, designers work in teams and must communicate to each other, the client, and likely the end user. Teamwork is important in engineering design because multiple perspectives and differing skills and knowledge are valuable when working to solve problems. Communication is key to the success of the designed solution. Designers must communicate their ideas clearly using many different representations, such as text in an engineering notebook, diagrams, flowcharts, technical briefs, or memos to the client.

LEARNING CYCLE

The same format for the learning cycle is used in all grade levels throughout the STEM Road Map, so that students engage in a variety of activities to learn about phenomena in the modules thoroughly and have consistent experiences in the problem-and project-based learning modules. Expectations for learning by younger students are

not as high as for older students, but the format of the progression of learning is the same. Students who have learned with curriculum from the STEM Road Map in early grades know what to expect in later grades. The learning cycle consists of five parts – Introductory Activity/Engagement, Activity/Exploration, Explanation, Elaboration/Application of Knowledge, and Evaluation/Assessment – and is based on the empirically tested 5E model from BSCS (Bybee et al. 2006).

In the Introductory Activity/Engagement phase, teachers introduce the module challenge and use a unique approach designed to pique students' curiosity. This phase gets students to start thinking about what they already know about the topic and begin wondering about key ideas. The Introductory Activity/Engagement phase positions students to be confident about what they are about to learn, because they have prior knowledge, and clues them into what they don't yet know.

In the Activity/Exploration phase, the teacher sets up activities in which students experience a deeper look at the topics that were introduced earlier. Students engage in the activities and generate new questions or consider possibilities using preliminary investigations. Students work independently, in small groups, and in whole-group settings to conduct investigations, resulting in common experiences about the topic and skills involved in the real-world activities. Teachers can assess students' development of concepts and skills based on the common experiences during this phase.

During the Explanation phase, teachers direct students' attention to concepts they need to understand and skills they need to possess to accomplish the challenge. Students participate in activities to demonstrate their knowledge and skills to this point, and teachers can pinpoint gaps in student knowledge during this phase.

In the Elaboration/Application of Knowledge phase, teachers present students with activities that engage in higher-order thinking to create depth and breadth of student knowledge, while connecting ideas across topics within and across STEM. Students apply what they have learned thus far in the module to a new context or elaborate on what they have learned about the topic to a deeper level of detail.

In the last phase, Evaluation/Assessment, teachers give students summative feedback on their knowledge and skills as demonstrated through the challenge. This is not the only point of assessment (as discussed in the section on Embedded Formative Assessments), but it is an assessment of the culmination of the knowledge and skills for the module. Students demonstrate their cognitive growth at this point and reflect on how far they have come since the beginning of the module. The challenges are designed to be multidimensional in the ways students must collaborate and communicate their new knowledge.

STEM RESEARCH NOTEBOOK

One of the main components of the *STEM Road Map Curriculum Series* is the STEM Research Notebook, a place for students to capture their ideas, questions, observations,

reflections, evidence of progress, and other items associated with their daily work. At the beginning of each module, the teacher walks students through the setup of the STEM Research Notebook, which could be a three-ring binder, composition book, or spiral notebook. You may wish to have students create divided sections so that they can easily access work from various disciplines during the module. Electronic notebooks kept on student devices are also acceptable and encouraged. Students will develop their own table of contents and create chapters in the notebook for each module.

Each lesson in the *STEM Road Map Curriculum Series* includes one or more prompts that are designed for inclusion in the STEM Research Notebook and appear as questions or statements that the teacher assigns to students. These prompts require students to apply what they have learned across the lesson to solve the big problem or challenge for that module. Each lesson is designed to meaningfully refer students to the larger problem or challenge they have been assigned to solve with their teams. The STEM Research Notebook is designed to be a key formative assessment tool, as students' daily entries provide evidence of what they are learning. The notebook can be used as a mechanism for dialogue between the teacher and students, as well as for peer and self-evaluation.

The use of the STEM Research Notebook is designed to scaffold student notebooking skills across the grade bands in the *STEM Road Map Curriculum Series*. In the early grades, children learn how to organize their daily work in the notebook as a way to collect their products for future reference. In elementary school, students structure their notebooks to integrate background research along with their daily work and lesson prompts. In the upper grades (middle and high school), students expand their use of research and data gathering through team discussions to more closely mirror the work of STEM experts in the real world.

THE ROLE OF ASSESSMENT IN THE *STEM ROAD MAP CURRICULUM SERIES*

Starting in the middle years and continuing into secondary education, the word *assessment* typically brings grades to mind. These grades may take the form of a letter or a percentage, but they typically are used as a representation of a student's content mastery. If well thought out and implemented, however, classroom assessment can offer teachers, parents, and students valuable information about student learning and misconceptions that does not necessarily come in the form of a grade (Popham 2013).

The *STEM Road Map Curriculum Series* provides a set of assessments for each module. Teachers are encouraged to use assessment information for more than just assigning grades to students. Instead, assessments of activities requiring students to actively engage in their learning, such as student journaling in STEM Research Notebooks, collaborative presentations, and constructing graphic organizers, should be used to move student learning forward. Whereas other curriculum with assessments may include

objective-type (multiple-choice or matching) tests, quizzes, or worksheets, we have intentionally avoided these forms of assessments to better align assessment strategies with teacher instruction and student learning techniques. Since the focus of this book is on project-or problem-based STEM curriculum and instruction that focuses on higher-level thinking skills, appropriate and authentic performance assessments were developed to elicit the most reliable and valid indication of growth in student abilities (Brookhart and Nitko 2008).

Comprehensive Assessment System

Assessment throughout all STEM Road Map curriculum modules acts as a comprehensive system in which formative and summative assessments work together to provide teachers with high-quality information on student learning. Formative assessment occurs when the teacher finds out formally or informally what a student knows about a smaller, defined concept or skill and provides timely feedback to the student about his or her level of proficiency. Summative assessments occur when students have performed all activities in the module and are given a cumulative performance evaluation in which they demonstrate their growth in learning.

A comprehensive assessment system can be thought of as akin to a sporting event. Formative assessments are the practices: It is important to accomplish them consistently, they provide feedback to help students improve their learning, and making mistakes can be worthwhile if students are given an opportunity to learn from them. Summative assessments are the competitions: Students need to be prepared to perform at the best of their ability. Without multiple opportunities to practice skills along the way through formative assessments, students will not have the best chance of demonstrating growth in abilities through summative assessments (Black and Wiliam 1998).

Embedded Formative Assessments

Formative assessments in this module serve two main purposes: to provide feedback to students about their learning and to provide important information for the teacher to inform immediate instructional needs. Providing feedback to students is particularly important when conducting problem-or project-based learning because students take on much of the responsibility for learning, and teachers must facilitate student learning in an informed way. For example, if students are required to conduct research for the Activity/Exploration phase but are not familiar with what constitutes a reliable resource, they may develop misconceptions based on poor information. When a teacher monitors this learning through formative assessments and provides specific feedback related to the instructional goals, students are less likely to develop incomplete or incorrect conceptions in their independent investigations. By using formative assessment to detect problems in student learning and then acting on this information, teachers help move student learning forward through these teachable moments.

Formative assessments come in a variety of formats. They can be informal, such as asking students probing questions related to student knowledge or tasks or simply observing students engaged in an activity to gather information about student skills. Formative assessments can also be formal, such as a written quiz or a laboratory practical. Regardless of the type, three key steps must be completed when using formative assessments (Sondergeld, Bell, and Leusner 2010). First, the assessment is delivered to students so that teachers can collect data. Next, teachers analyze the data (student responses) to determine student strengths and areas that need additional support. Finally, teachers use the results from information collected to modify lessons and create learning environments that reinforce weak points in student learning. If student learning information is not used to modify instruction, the assessment cannot be considered formative in nature. Formative assessments can be about content, science process skills, or even learning skills. When a formative assessment focuses on content, it assesses student knowledge about the disciplinary core ideas from the *Next Generation Science Standards* (*NGSS*) or content objectives from *Common Core State Standards for Mathematics* (*CCSS Mathematics*) or *Common Core State Standards for English Language Arts* (*CCSS ELA*). Content-focused formative assessments ask students questions about declarative knowledge regarding the concepts they have been learning. Process skills formative assessments examine the extent to which a student can perform science and engineering practices from the *NGSS* or process objectives from *CCSS Mathematics* or *CCSS ELA*, such as constructing an argument. Learning skills can also be assessed formatively by asking students to reflect on the ways they learn best during a module and identify ways they could have learned more.

Assessment Maps

Assessment maps or blueprints can be used to ensure alignment between classroom instruction and assessment. If what students are learning in the classroom is not the same as the content on which they are assessed, the resultant judgment made on student learning will be invalid (Brookhart and Nitko 2008). Therefore, the issue of instruction and assessment alignment is critical. The assessment map for this book (found in Chapter 3) indicates by lesson whether the assessment should be completed as a group or on an individual basis, identifies the assessment as formative or summative in nature, and aligns the assessment with its corresponding learning objectives.

Note that the module includes far more formative assessments than summative assessments. This is done intentionally to provide students with multiple opportunities to practice their learning of new skills before completing a summative assessment. Note also that formative assessments are used to collect information on only one or two learning objectives at a time so that potential relearning or instructional modifications can focus on smaller and more manageable chunks of information. Conversely, summative assessments in the module cover many more learning objectives, as they are traditionally used as final markers of student learning. This is not to say that information collected from summative assessments cannot or should not be used formatively. If teachers find that gaps in student learning persist after a summative assessment is completed, it is important to revisit these existing misconceptions or areas of weakness before moving on (Black et al., 2003).

SELF-REGULATED LEARNING THEORY IN THE STEM ROAD MAP MODULES

Many learning theories are compatible with the STEM Road Map modules, such as constructivism, situated cognition, and meaningful learning. However, we feel that the self-regulated learning theory (SRL) aligns most appropriately (Zimmerman, 2000). SRL requires students to understand that thinking needs to be motivated and managed (Ritchhart, Church, and Morrison 2011). The STEM Road Map modules are student centered and are designed to provide students with choices, concrete hands-on experiences, and opportunities to see and make connections, especially across subjects (Eliason and Jenkins 2012; NAEYC 2016). Additionally, SRL is compatible with the modules because it fosters a learning environment that supports students' motivation, enables students to become aware of their own learning strategies, and requires reflection on learning while experiencing the module (Peters and Kitsantas 2010).

The theory behind SRL (see Figure 2.2) explains the different processes that students engage in before, during, and after a learning task. Because SRL is a cyclical learning process, the accomplishment of one cycle develops strategies for the next learning cycle. This cyclic way of learning aligns with the various sections in the STEM Road Map lesson plans on Introductory Activity/Engagement, Activity/Exploration, Explanation, Elaboration/Application of Knowledge, and Evaluation/Assessment. Since the students engaged in a module take on much of the responsibility for learning, this theory also provides guidance for teachers to keep students on the right track.

Figure 2.2. SRL Theory

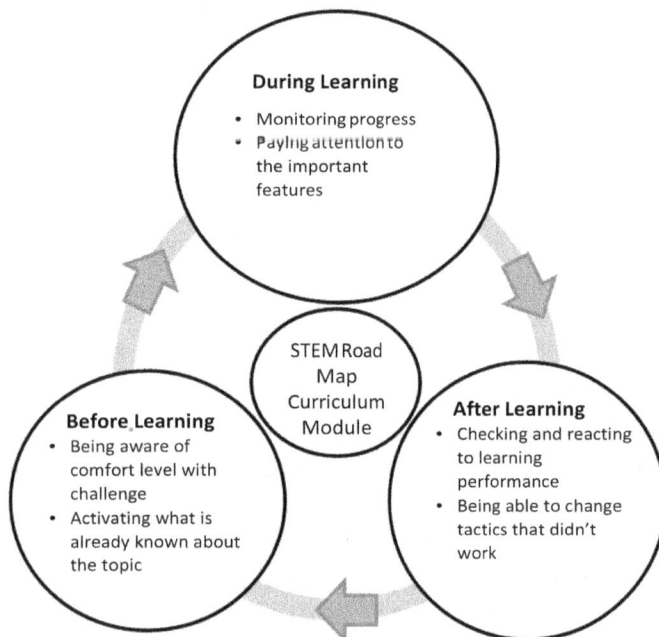

Source: Adapted from Zimmerman 2000.

The remainder of this section explains how SRL theory is embedded within the five sections of each module and points out ways to support students in becoming independent learners of STEM while productively functioning in collaborative teams.

Before Learning: Setting the Stage

Before attempting a learning task such as the STEM Road Map modules, teachers should develop an understanding of their students' level of comfort with the process of accomplishing the learning and determine what they already know about the topic. When students are comfortable with attempting a learning task, they tend to take more risks in learning and as a result achieve deeper learning (Bandura 1986).

The STEM Road Map curriculum modules are designed to foster excitement from the very beginning. Each module has an Introductory Activity/Engagement section that introduces the overall topic from a unique and exciting perspective, engaging the students to learn more so that they can accomplish the challenge. The Introductory Activity also has a design component that helps teachers assess what students already know about the topic of the module. In addition to the deliberate designs in the lesson plans to support SRL, teachers can support a high level of student comfort with the learning challenge by finding out if students have ever accomplished the same kind of task and, if so, asking them to share what worked well for them.

During Learning: Staying the Course

Some students fear inquiry learning because they aren't sure what to do to be successful (Peters, 2010). However, the STEM Road Map curriculum modules are embedded with tools to help students pay attention to knowledge and skills that are important for the learning task and to check student understanding along the way. One of the most important processes for learning is the ability for learners to monitor their own progress while performing a learning task (Peters 2012). The modules allow students to monitor their progress with tools such as the STEM Research Notebooks, in which they record what they know and can check whether they have acquired a complete set of knowledge and skills. The STEM Road Map modules support inquiry strategies that include previewing, questioning, predicting, clarifying, observing, discussing, and journaling (Morrison and Milner 2014). Through the use of technology throughout the modules, inquiry is supported by providing students access to resources and data while enabling them to process information, report the findings, collaborate, and develop 21st century skills.

It is important for teachers to encourage students to have an open mind about alternative solutions and procedures (Milner and Sondergeld 2015) when working through the STEM Road Map curriculum modules. Novice learners can have difficulty knowing what to pay attention to and tend to treat each possible avenue for information as equal (Benner 1984). Teachers are the mentors in a classroom and can point out ways

for students to approach learning during the Activity/Exploration, Explanation, and Elaboration/Application of Knowledge portions of the lesson plans to ensure that students pay attention to the important concepts and skills throughout the module. For example, if a student is to demonstrate conceptual awareness of motion when working on roller coaster research, but the student has misconceptions about motion, the teacher can step in and redirect student learning.

After Learning: Knowing What Works

The classroom is a busy place, and it may often seem that there is no time for self-reflection on learning. Although skipping this reflective process may save time in the short term, it reduces the ability to take into account things that worked well and things that didn't so that teaching the module may be improved next time. In the long run, SRL skills are critical for students to become independent learners who can adapt to new situations. By investing the time it takes to teach students SRL skills, teachers can save time later, because students will be able to apply methods and approaches for learning that they have found effective to new situations. In the Evaluation/Assessment portion of the STEM Road Map curriculum modules, as well as in the formative assessments throughout the modules, two processes in the after-learning phase are supported: evaluating one's own performance and accounting for ways to adapt tactics that didn't work well. Students have many opportunities to self-assess in formative assessments, both in groups and individually, using the rubrics provided in the modules.

The designs of the *NGSS* and *CCSS* allow for students to learn in diverse ways, and the STEM Road Map curriculum modules emphasize that students can use a variety of tactics to complete the learning process. For example, students can use STEM Research Notebooks to record what they have learned during the various research activities. Notebook entries might include putting objectives in students' own words, compiling their prior learning on the topic, documenting new learning, providing proof of what they learned, and reflecting on what they felt successful doing and what they felt they still needed to work on. Perhaps students didn't realize that they were supposed to connect what they already knew with what they learned. They could record this and would be prepared in the next learning task to begin connecting prior learning with new learning.

SAFETY IN STEM

Student safety is a primary consideration in all subjects but is an area of particular concern in science, where students may interact with unfamiliar tools and materials that may pose additional safety risks. It is important to implement safety practices within the context of STEM investigations, whether in a classroom laboratory or in the field. When you keep safety in mind as a teacher, you avoid many potential issues with the lesson while also protecting your students.

STEM safety practices encompass things considered in the typical science classroom. Ensure that students are familiar with basic safety considerations, such as wearing protective equipment (e.g., safety glasses or goggles and latex-free gloves) and taking care with sharp objects, and know emergency exit procedures. Teachers should learn beforehand the locations of the safety eyewash, fume hood, fire extinguishers, and emergency shut-off switch in the classroom and how to use them. Also be aware of any school or district safety policies that are in place and apply those that align with the work being conducted in the lesson. It is important to review all safety procedures annually.

STEM investigations should always be supervised. Each lesson in the modules includes teacher guidelines for applicable safety procedures that should be followed. Before each investigation, teachers should go over these safety procedures with the student teams. Some STEM focus areas such as engineering require that students can demonstrate how to properly use equipment in the maker space before the teacher allows them to proceed with the lesson.

Information about classroom science safety, including a safety checklist for science classrooms, general lab safety recommendations, and links to other science safety resources, is available at the Council of State Science Supervisors (CSSS) website at *www.csss-science.org/safety.shtml*. The National Science Teachers Association (NSTA) provides a list of science rules and regulations, including standard operating procedures for lab safety, and a safety acknowledgement form for students and parents or guardians to sign. You can access these resources at *http://static.nsta.org/pdfs/Safety InTheScienceClassroom.pdf*. In addition, NSTA's Safety in the Science Classroom web page *(www.nsta.org/safety)* has numerous links to safety resources, including papers written by the NSTA Safety Advisory Board.

Disclaimer: The safety precautions for each activity are based on use of the recommended materials and instructions, legal safety standards, and better professional practices. Using alternative materials or procedures for these activities may jeopardize the level of safety and therefore is at the user's own risk.

REFERENCES

Bandura, A. 1986. *Social foundations of thought and action: A social cognitive theory*. Englewood Cliffs, NJ: Prentice-Hall.

Barell, J. 2006. *Problem-based learning: An inquiry approach*. Thousand Oaks, CA: Corwin Press.

Benner, P. 1984. *From novice to expert: Excellence and power in clinical nursing practice*. Menlo Park, CA: Addison-Wesley Publishing Company.

Black, P., C. Harrison, C. Lee, B. Marshall, and D. Wiliam. 2003. *Assessment for learning: Putting it into practice*. Berkshire, UK: Open University Press.

Black, P., and D. Wiliam. 1998. Inside the black box: Raising standards through classroom assessment. *Phi Delta Kappan* 80 (2): 139–148.

Blumenfeld, P., E. Soloway, R. Marx, J. Krajcik, M. Guzdial, and A. Palincsar. 1991. Motivating project-based learning: Sustaining the doing, supporting learning. *Educational Psychologist* 26 (3): 369–398.

Brookhart, S. M., and A. J. Nitko. 2008. *Assessment and grading in classrooms.* Upper Saddle River, NJ: Pearson.

Bybee, R., J. Taylor, A. Gardner, P. Van Scotter, J. Carlson, A. Westbrook, and N. Landes. 2006. *The BSCS 5E instructional model: Origins and effectiveness. https://bscs.org/wp-content/uploads/2022/01/bscs_5e_full_report-1.pdf.*

Eliason, C. F., and L. T. Jenkins. 2012. *A practical guide to early childhood curriculum.* 9th ed. New York: Merrill.

Johnson, C. 2003. Bioterrorism is real-world science: Inquiry-based simulation mirrors real life. *Science Scope* 27 (3): 19–23.

Krajcik, J., and P. Blumenfeld. 2006. Project-based learning. In *The Cambridge handbook of the learning sciences,* ed. R. Keith Sawyer, 317–334. New York: Cambridge University Press.

Lambros, A. 2004. *Problem-based learning in middle and high school classrooms: A teacher's guide to implementation.* Thousand Oaks, CA: Corwin Press.

Milner, A. R., and T. Sondergeld. 2015. Gifted urban middle school students: The inquiry continuum and the nature of science. *National Journal of Urban Education and Practice* 8 (3): 442–461.

Morrison, V., and A. R. Milner. 2014. Literacy in support of science: A closer look at cross-curricular instructional practice. *Michigan Reading Journal* 46 (2): 42–56.

National Association for the Education of Young Children (NAEYC). 2016. Developmentally appropriate practice position statements. *www.naeyc.org/positionstatements/dap.*

Peters, E. E. 2010. Shifting to a student-centered science classroom: An exploration of teacher and student changes in perceptions and practices. *Journal of Science Teacher Education* 21 (3): 329–349.

Peters, E. E. 2012. Developing content knowledge in students through explicit teaching of the nature of science: Influences of goal setting and self-monitoring. *Science and Education* 21 (6): 881–898.

Peters, E. E., and A. Kitsantas. 2010. The effect of nature of science metacognitive prompts on science students' content and nature of science knowledge, metacognition, and self-regulatory efficacy. *School Science and Mathematics* 110: 382–396.

Popham, W. J. 2013. *Classroom assessment: What teachers need to know.* 7th ed. Upper Saddle River, NJ: Pearson.

Ritchhart, R., M. Church, and K. Morrison. 2011. *Making thinking visible: How to promote engagement, understanding, and independence for all learners.* San Francisco, CA: Jossey-Bass.

Sondergeld, T. A., C. A. Bell, and D. M. Leusner. 2010. Understanding how teachers engage in formative assessment. *Teaching and Learning* 24 (2): 72–86.

Zimmerman, B. J. 2000. Attaining self-regulation: A social-cognitive perspective. In *Handbook of self-regulation,* ed. M. Boekaerts, P. Pintrich, and M. Zeidner, 13–39. San Diego: Academic Press.

PART 2

HABITATS LOCAL AND FAR AWAY

STEM ROAD MAP MODULE

HABITATS LOCAL AND FAR AWAY MODULE OVERVIEW

*Andrea R. Milner, Vanessa B. Morrison, Janet B. Walton,
Carla C. Johnson, and Erin E. Peters-Burton*

THEME: Sustainable Systems

LEAD DISCIPLINES: Science and Social Studies

MODULE SUMMARY

In this module, students will explore habitats on a global scale. Student teams are each challenged to learn about an endangered species and develop an action plan designed to encourage others to work to save their team's chosen species. Students will first explore local habitats and the needs of living things, and then explore endangered species and habitat connections at the local level. Next, students will use a global lens to investigate how habitat loss impacts species by investigating globally endangered species. Finally, teams will create action plans that describe one endangered species' habitat characteristics and why the species is endangered, and will offer solutions about how humans might be able to support this species' survival. At the end of the module, student teams will present their action plans (adapted from Koehler, Bloom, and Milner, 2015).

ESTABLISHED GOALS AND OBJECTIVES

At the conclusion of this module, students will be able to do the following:

- Explain that there are various types of habitats that vary with geographical location around the world

- Identify several habitats in the U.S. and globally

- Explain how various habitats meet animals' basic needs

- Identify climatic characteristics of several habitats

- Identify several endangered species

- Identify several examples of interdependence in ecosystems

- Identify humans as species that live within and in interaction with various habitats

- Identify several ways that humans can impact habitats

- Design and construct models to demonstrate understanding of features of various habitats (local and global) and endangered species

- Apply their knowledge of habitat characteristics, interdependence in ecosystems, and endangered species to develop an action plan to help preserve an endangered species

CHALLENGE OR PROBLEM FOR STUDENTS TO SOLVE: THE SAVE THE SPECIES CHALLENGE

In this module, students will act as explorers as they investigate animals locally and around the world. Student teams will each be challenged to choose an endangered species in a habitat outside of the U.S. (or in a North American region outside of their home region) and develop an action plan describing the climate, the animals and plants that reside in the species' habitat, and how humans and other factors may influence this habitat and contribute to its vitality or pose challenges to it.

CONTENT STANDARDS ADDRESSED IN THIS STEM ROAD MAP MODULE

A full listing with descriptions of the standards this module addresses can be found in Appendix C. Listings of the particular standards addressed within lessons are provided in a table for each lesson in Chapter 4.

STEM RESEARCH NOTEBOOK

Each student should maintain a STEM Research Notebook, which will serve as a place for students to organize their work throughout this module (see pp. 12–13 for more general discussion on setup and use of this notebook). All written work in the module should be included in the notebook, including records of students' thoughts and ideas, fictional accounts based on the concepts in the module, and records of student progress through the engineering design process (EDP). The notebooks may be maintained across subject areas, giving students the opportunity to see that although their classes may be separated during the school day, the knowledge they gain is connected. Templates for the STEM Research Notebook pages for this module are included in Appendix A.

Emphasize to students the importance of organizing all information in a Research Notebook. Explain to them that scientists and other researchers maintain detailed Research Notebooks in their work. These notebooks, which are crucial to researchers' work because they contain critical information and track the researchers' progress, are often considered legal documents for scientists who are pursuing patents or wish to provide proof of their discovery process.

MODULE LAUNCH

Lesson Plan 1 (see Chapter 4, p. 39) will begin by launching the module. You will hold a class discussion about endangered species and habitats, asking students these questions:

- What is a species?

- What do we mean by endangered species?

- Are there different types of endangered species?

- Why might an animal species be endangered?

- Can a plant be an endangered species?

- Where and when have you seen endangered species?

As a class, explore endangered species by watching a video about endangered species such as "Top 10 Most Endangered Species" at *www.youtube.com/watch?v=fm8q TACshos.*

PREREQUISITE SKILLS FOR THE MODULE

Students enter this module with a wide range of preexisting skills, information, and knowledge. Table 3.1 provides an overview of prerequisite skills and knowledge

Table 3.1. Prerequisite Key Knowledge and Examples of Applications and Differentiation

Prerequisite Key Knowledge	Application of Knowledge	Differentiation for Students Needing Knowledge
Science		
• Cause and effect	• Determine how components of various habitats such as climate and animal and plant life are interdependent. • Determine how human behavior can influence habitats and species that live in those habitats.	• Provide demonstrations of cause and effect (e.g., dropping egg [cause] and observing breakage [effect]), emphasizing that cause is why something happens, effect is what happens. • Read aloud picture books to class and have students identify cause and effect sequences. • Create a class T-chart to record causes and related effects students observe in the classroom and in literature.

continued

Table 3.1. (*continued*)

Prerequisite Key Knowledge	Application of Knowledge	Differentiation for Students Needing Knowledge
Mathematics		
• Number sense	• Measure characteristics of their local habitat. • Describe components of habitats numerically.	• Model measurement techniques using standard and nonstandard units of measurement. • Read aloud nonfiction texts about measurement to class. • Provide opportunities for students to practice measurement and counting in a variety of settings (e.g., in the classroom and outdoors).
Language and Inquiry Skills		
• Visualize • Make predictions • Ask and respond to questions	• Make and confirm or reject predictions. • Share thought processes through notebooking, asking and responding to questions, and using the EDP.	• As a class, make predictions when reading fictional texts. • Model the process of using information and prior knowledge to make predictions. • Provide samples of notebook entries.
Speaking and Listening		
• Participate in group discussions	• Engage in collaborative group discussions in the development of their reference guide and in completing group tasks throughout the module.	• Model speaking and listening skills. • Create a class list of good listening and good speaking practices. • Read picture books that feature collaboration and teamwork.

that students are expected to apply in this module, along with examples of how they apply this knowledge throughout the module. Differentiation strategies are also provided for students who may need additional support in acquiring or applying this knowledge.

POTENTIAL STEM MISCONCEPTIONS

Students enter the classroom with a wide variety of prior knowledge and ideas, so it is important to be alert to misconceptions or inappropriate understandings of foundational knowledge. These misconceptions can be classified as one of several types: "preconceived notions," opinions based on popular beliefs or understandings; "nonscientific beliefs," knowledge students have gained about science from sources outside the scientific community; "conceptual misunderstandings," incorrect conceptual models based on incomplete understanding of concepts; "vernacular misconceptions," misunderstandings of words based on their common use versus their scientific use; and "factual misconceptions," incorrect or imprecise knowledge learned in early life that remains unchallenged (NRC 1997, p. 28). Misconceptions must be addressed and dismantled in order for students to reconstruct their knowledge, and therefore teachers should be prepared to take the following steps:

- Identify students' misconceptions.

- Provide a forum for students to confront their misconceptions.

- Help students reconstruct and internalize their knowledge, based on scientific models.

(NRC 1997, p. 29)

Keeley and Harrington (2010) recommend using diagnostic tools such as probes and formative assessment to identify and confront student misconceptions and begin the process of reconstructing student knowledge. Keeley and Harrington's *Uncovering Student Ideas in Science* series contains probes targeted toward uncovering student misconceptions in a variety of areas and may be a useful resource for addressing student misconceptions in this module. In addition, Know, Want to Know, Learned (KWL) charts are used throughout this module. These charts are completed as a class, and will provide useful information about students' existing knowledge and misconceptions regarding lesson concepts.

Some commonly held misconceptions specific to lesson content are provided with each lesson so that you can be alert for student misunderstanding of the science concepts presented and used during this module.

SELF-REGULATED LEARNING (SRL) PROCESS COMPONENTS

Table 3.2 illustrates some of the activities in the Habitats Local and Far Away module and how they align to the SRL processes before, during, and after learning. See Chapter 2 (pp. 16–18) for an overview of SRL theory.

Table 3.2. SRL Learning Components

Learning Process Components	Example from Habitats Local and Far Away Module	Lesson Number and Learning Component
Before Learning		
Motivates students	Students make predictions about what animals are endangered before watching a video on the subject.	Lesson 1 Introductory Activity/ Engagement
Evokes prior learning	Students take a walking tour of the school neighborhood to identify local habitats.	Lesson 1 Activity/Investigation
During Learning		
Focuses on important features	In the Habitat Helpers activity, students identify some of the reasons species locally are threatened. Students share ideas and the teacher directs students' attention to the important features.	Lesson 2 Activity/Investigation
Helps students monitor their progress	Students create a simulated habitat for their local endangered species. Using the EDP, students define, learn, plan, try, test, and decide on their model's design. During the test phase, students decide whether to proceed with their current choices or change direction based on feedback.	Lesson 2 Activity/Investigation
After Learning		
Evaluates learning	Students present action plans, describing the species' habitat, reasons for its endangered status, and actions humans can take to support the species' survival. The action plans can be video recorded to allow students to self-evaluate; classroom guests or other audience members also provide feedback to the groups.	Lesson 4 Elaboration/Application of Knowledge
Takes account of what worked and what did not work	The whole class discusses and analyses strengths of each group's challenge product and suggests improvements. Presenting groups respond with ways that they could adapt their plan based on the discussion.	Lesson 4 Elaboration/Application of Knowledge

STRATEGIES FOR DIFFERENTIATING INSTRUCTION WITHIN THIS MODULE

For the purposes of this curriculum module, differentiated instruction is conceptualized as a way to tailor instruction – including process, content, and product – to various student needs in your class. A number of differentiation strategies are integrated into lessons across the module. The problem- and project-based learning approach used in the lessons is designed to address students' multiple intelligences by providing a variety of entry points and methods to investigate the key concepts in the module (for example, investigating habitats via scientific inquiry, literature, journaling, and collaborative design). Differentiation strategies for students needing support in prerequisite knowledge can be found in Table 3.1 (pp. 25–26). You are encouraged to use information gained about student prior knowledge during introductory activities and discussions to inform your instructional differentiation. Strategies incorporated into this lesson include flexible grouping, varied environmental learning contexts, assessments, compacting, tiered assignments and scaffolding, and mentoring. The following websites may be helpful resources for differentiated instruction:

- *https://steinhardt.nyu.edu/metrocenter/research-evaluation/research/culturally-responsive-sustaining-education*

- *http://educationnorthwest.org/sites/default/files/12.99.pdf*

Flexible Grouping: Students work collaboratively in a variety of activities throughout this module. Grouping strategies you might employ include student-led grouping, grouping students according to ability level or common interests, grouping students randomly, or grouping them so that students in each group have complementary strengths (for instance, one student might be strong in mathematics, another in art, and another in writing).

Varied Environmental Learning Contexts: Students have the opportunity to learn in various contexts throughout the module, including alone, in groups, in quiet reading and research-oriented activities, and in active learning through inquiry and design activities. In addition, students learn in a variety of ways, including through doing inquiry activities, journaling, reading a variety of texts, watching videos, participating in class discussion, and conducting web-based research.

Assessments: Students are assessed in a variety of ways throughout the module, including individual and collaborative formative and summative assessments. Students have the opportunity to produce work via written text, oral presentations, and modeling.

Compacting: Based on student prior knowledge, you may wish to adjust instructional activities for students who exhibit prior mastery of a learning objective. Since student work in science is largely collaborative throughout the module, this

strategy may be most appropriate for mathematics, social studies, or English language arts activities.

Tiered Assignments and Scaffolding: Based on your awareness of student ability, understanding of concepts, and mastery of skills, you may wish to provide students with variations on activities by adding complexity to assignments or providing more or fewer learning supports for activities throughout the module. For instance, some students may need additional support in identifying key search words and phrases for web-based research or may benefit from cloze sentence handouts to enhance vocabulary understanding.

Other students may benefit from expanded reading selections and additional reflective writing or from working with manipulatives and other visual representations of mathematical concepts. You may also work with your school librarian to compile a classroom database of research resources and supplementary readings for different reading levels and on a variety of topics related to the module challenge to provide opportunities for students to undertake independent reading. You may find the following website on scaffolding strategies helpful: *www.edutopia.org/blog/ scaffolding-lessons-six-strategies-rebecca-alber.*

Mentoring: As group design teamwork becomes increasingly complex throughout the module, you may wish to have a resource teacher, older student, or volunteer work with groups that struggle to stay on task and collaborate effectively.

STRATEGIES FOR ENGLISH LANGUAGE LEARNERS (ELLS)

Students who are developing proficiency in English language skills require additional supports to simultaneously learn academic content and the specialized language associated with specific content areas. WIDA has created a framework for providing support to these students and makes available rubrics and guidance on differentiating instructional materials for English language learners (ELLs) (see *www.wida.us*). In particular, ELL students may benefit from additional sensory supports such as images, physical modeling, and graphic representations of module content, as well as interactive support through collaborative work. This module incorporates a variety of sensory supports and offers ongoing opportunities for ELL students to work collaboratively.

Teachers differentiating instruction for multilingual learners should carefully consider the needs of these students as they introduce and use academic language in various language domains (listening, speaking, reading, and writing) throughout this module. To adequately differentiate instruction for ELL students, teachers should have an understanding of the proficiency level of each student. The following five overarching preK–5 WIDA learning standards are relevant to this module:

- Standard 1: Social and Instructional Language. Focus on following directions, personal information, collaboration with peers.

- Standard 2: The language of Language Arts. Focus on nonfiction, fiction, sequence of story, elements of story.

- Standard 3: The language of Mathematics. Focus on basic operations, number sense, interpretation of data, patterns.

- Standard 4: The language of Science. Focus on forces in nature, scientific process, living and nonliving things, organisms, and environment.

- Standard 5: The language of Social Studies. Focus on homes and habitats, jobs and careers, geography, representations of Earth (maps and globes).

SAFETY CONSIDERATIONS FOR THE ACTIVITIES IN THIS MODULE

Science activities in this module focus on exploring habitats and the living things found in habitats. Students will work with a variety of materials as they explore animals and habitats. You should discuss appropriate use of materials with students at the start of each activity. For more general safety guidelines, see the Safety in STEM section in Chapter 2 (pp. 18–19) and for lesson-specific safety information, see the Safety Notes section of each lesson in Chapter 4.

DESIRED OUTCOMES AND MONITORING SUCCESS

The desired outcomes for this module are outlined in Table 3.3, along with suggested ways to gather evidence to monitor student success. For more specific details on

Table 3.3. Desired Outcomes and Evidence of Success in Achieving Identified Outcomes

Desired Outcome	Evidence of Success in Achieving Identified Outcome	
	Performance Tasks	Other Measures
Students will understand and demonstrate their knowledge about life in places near and far by comparing and contrasting various habitats (local and global) and will also demonstrate their understanding of interdependence in ecosystems.	• Student teams will develop and present action plans to propose ways to ensure the survival of an endangered species. The plan will describe habitat characteristics, explain why the species is endangered, and offer solutions about how humans might be able to support this species' survival.	Students are assessed using the Observation, STEM Research Notebook, and Participation Rubric.

desired outcomes, see the Established Goals and Objectives sections for the module and individual lessons.

ASSESSMENT PLAN OVERVIEW AND MAP

Table 3.4 provides an overview of the major group and individual *products* and *deliverables*, or things that student teams will produce in this module, that constitute the assessment for this module. See Table 3.5 (pp. 32–34) for a full assessment map of formative and summative assessments in this module.

Table 3.4. Major Products and Deliverables in Lead Disciplines for Groups and Individuals

Lesson	Major Group Products/ Deliverables	Major Individual Products/ Deliverables
1	• Neighborhood Explorers presentations	• STEM Research Notebook entries 1–7 • Lesson Assessment
2	• Mighty Mobiles and presentations • Habitat Helpers models and presentations	• STEM Research Notebook entries 8–12 • Lesson Assessment
3	• Amazing Adaptations models and presentations	• STEM Research Notebook entries 9–17 • Lesson Assessment
4	• Team research and poster for class conservation plan for local endangered species • Save the Species Action Plan presentations	• Save the Species Action Plan

Table 3.5. Assessment Map for Habitats Local and Far Away Module

Lesson	Assessment	Group/ Individual	Formative/ Summative	Lesson Objective Assessed
1	STEM Research Notebook *entries*	Individual	Formative	• Identify several habitats and their characteristics. • Identify the basic needs of living things. • Describe how habitats provide living things with their basic needs.
1	Neighborhood Explorers *performance task*	Group	Formative	• Observe and describe features of local habitats. • Describe how habitats provide living things with their basic needs.

Lesson	Assessment	Group/ Individual	Formative/ Summative	Lesson Objective Assessed
1	Lesson *assessment*	Individual	Summative	• Identify the basic needs of living things. • Describe how habitats provide living things with their basic needs.
2	STEM Research Notebook *entries*	Individual	Formative	• Identify local endangered animal species. • Identify several reasons for animal species becoming endangered. • Identify ways that humans can help to prevent endangered species from becoming extinct. • Identify engineering as a career and describe the kind of work engineers do. • Identify several examples of interdependency in ecosystems. • Depict a simple food chain using words and drawings.
2	Mighty Mobiles *performance task*	Group	Formative	• Identify local endangered animal species. • Identify several reasons for animal species becoming endangered. • Identify ways that humans can help to prevent endangered species from becoming extinct. • Design and construct mobiles depicting local endangered animal species.
2	Habitat Helpers *model and presentation*	Group	Formative	• Identify local endangered animal species. • Use the steps of the EDP to create a model habitat for a local endangered animal species.
2	Lesson *assessment*	Individual	Summative	• Identify local endangered animal species. • Identify several reasons for animal species becoming endangered. • Identify ways that humans can help to prevent endangered species from becoming extinct. • Identify several examples of interdependency in ecosystems.
3	STEM Research Notebook *entries*	Individual	Formative	• Identify and describe several animal physical adaptations. • Design and construct models to reflect how animals use their external parts to help them survive, grow, and meet their needs. • Identify endangered species from other parts of the world. • Identify several reasons why species become endangered in other parts of the world.

continued

Table 3.5. (*continued*)

Lesson	Assessment	Group/ Individual	Formative/ Summative	Lesson Objective Assessed
3	Amazing Adaptations *models and presentations*	Group	Formative	• Identify endangered species from other parts of the world. • Identify several reasons why species become endangered in other parts of the world. • Identify and describe several animal physical adaptations. • Use the EDP to design and construct models to reflect how animals use their external parts to help them survive, grow, and meet their needs.
3	Lesson *assessment*	Individual	Summative	• Identify endangered species from other parts of the world. • Identify the continents and locate them on a map.
4	Class Conservation Plan *posters*	Group	Formative	• Apply their learning about habitats and endangered animals to create an action plan that describes habitats and an endangered species' interaction with that habitat. • Apply their learning about habitats and endangered animals to identify actions that people can take to ensure the survival of endangered species.
4	Save the Species Action *Plan and presentation*	Individual/ Group	Summative	• Apply their learning about habitats and endangered animals to create an action plan that describes habitats and an endangered species' interaction with that habitat. • Apply their learning about habitats and endangered animals to identify actions that people can take to ensure the survival of endangered species.

MODULE TIMELINE

Tables 3.6–3.10 provide lesson timelines for each week of the module. These timelines are provided for general guidance only and are based on class times of approximately 30 minutes.

Table 3.6. STEM Road Map Module Schedule Week 1

Day 1	Day 2	Day 3	Day 4	Day 5
Lesson 1 Earth's Amazing Endangered Species	*Lesson 1 Earth's Amazing Endangered Species*	*Lesson 1 Earth's Amazing Endangered Species*	*Lesson 1 Earth's Amazing Endangered Species*	*Lesson 1 Earth's Amazing Endangered Species*
Launch the module with a discussion of endangered species.	Discussion of habitats and continents.	Discussion about cultural and societal implications of living in different places.	Interactive read aloud of *Me and the Measure of Things* by Joan Sweeney.	Complete Neighborhood Explorers activity and presentations.
Show video about endangered species.	Interactive read aloud of *Nature's Patchwork Quilt: Understanding Habitats* by Mary Miche.	Introduce the five basic needs of living things.	Introduce scientific tools.	Interactive read aloud of *Counting on Frank* by Rod Clement.
Students draw and label an endangered species.	Discuss mathematical descriptions of habitats.	Students participate in the Five Needs game.	Students participate in the Neighborhood Explorers activity.	View video about reasons for animals becoming endangered.
	Start vocabulary chart.			

Table 3.7. STEM Road Map Module Schedule Week 2

Day 6	Day 7	Day 8	Day 9	Day 10
Lesson 1 Earth's Amazing Endangered Species	*Lesson 2 Our Local Endangered Species*	*Lesson 2 Our Local Endangered Species*	*Lesson 2 Our Local Endangered Species*	*Lesson 2 Our Local Endangered Species*
Lesson Assessment.	Introduce and discuss human conservation efforts.	Discussion about physical adaptations.	Finish Mighty Mobiles activity and have teams present their mobiles.	Introduce EDP and begin Habitat Helpers activity.
Discuss how endangered species are described mathematically.	Introduce American bison as an example of conservation success via a video and an interactive read aloud of *The Buffalo are Back* by Jean Craighead George.	Begin Mighty Mobiles activity.		
	Compare and contrast local habitat and bisons' habitat.			

Table 3.8. STEM Road Map Module Schedule Week 3

Day 11	Day 12	Day 13	Day 14	Day 15
Lesson 2 Our Local Endangered Species	*Lesson 2 Our Local Endangered Species*	*Lesson 2 Our Local Endangered Species*	*Lesson 3 Endangered Species Throughout the World*	*Lesson 3 Endangered Species Throughout the World*
Continue Habitat Helpers activity.	Conclude Habitat Helpers activity and have teams present their models.	Introduce interdependency.	Discussion about adaptations.	Begin Amazing Adaptations investigation.
Students explore engineering through an interactive read aloud of *Rosie Revere, Engineer* by Andrea Beaty.		Conduct interactive read aloud of *Who Eats What? Food Chains and Food Webs* by Patricia Lauber.	Review continents and match endangered species to continents.	
		Lesson Assessment.	Interactive read aloud of *What Do You Do with a Tail Like This?* by Steve Jenkins.	

3

Table 3.9. STEM Road Map Module Schedule Week 4

Day 16	Day 17	Day 18	Day 19	Day 20
Lesson 3 Endangered Species Throughout the World	*Lesson 3 Endangered Species Throughout the World!*	*Lesson 3 Endangered Species Throughout the World*	*Lesson 3 Endangered Species Throughout the World*	*Let's Explore Endangered Species throughout the World!*
Continue Amazing Adaptations investigation.	Continue Amazing Adaptations investigation.	Amazing Adaptations presentations.	Conduct an interactive read aloud from book choices. Discuss mathematical descriptions of animals. Review Point Arena Mountain Beaver recovery plan and brainstorm ideas for action plan.	Interactive read aloud of *Twelve Snails to One Lizard* by Susan Hightower. Assessment.

Table 3.10. STEM Road Map Module Schedule Week Five

Day 21	Day 22	Day 23	Day 24	Day 25
Lesson 4 The Save the Species Challenge	*Lesson 4 The Save the Species Challenge*	*Lesson 4 The Save the Species Challenge*	*Lesson 4 The Save the Species Challenge*	*Lesson 4 The Save the Species Challenge*
Class discussion about conservation. Begin Class Conservation Plan.	Complete Class Conservation Plan. Begin work on Save the Species Action Plans.	Continue work on Save the Species Action Plans.	Continue work on Save the Species Action Plans.	Student teams present their Save the Species Action Plans.

RESOURCES

The media specialist can help teachers locate resources for students to view and read about plants, weather, and related content. Special educators and reading specialists can help find supplemental sources for students needing extra support in reading and writing. Additional resources may be found online. Community resources for this module may include biologists, botanists, and zoologists.

REFERENCES

Keeley, P., and R. Harrington. 2010. *Uncovering student ideas in physical science, volume 1: 45 new force and motion assessment probes.* Arlington, VA: NSTA Press.

Koehler, C., M. A. Bloom, and A. R. Milner. 2015. The STEM Road Map for grades K–2. In *STEM Road Map: A framework for integrated STEM education,* ed. C. C. Johnson, E. E. Peters-Burton, and T. J. Moore, 41–67. New York: Routledge. *www.routledge.com/products/9781138804234.*

National Research Council (NRC). 1997. *Science teaching reconsidered: A handbook.* Washington, DC: National Academies Press.

WIDA. (2020). *WIDA English language development standards framework, 2020 edition: Kindergarten–grade 12.* Board of Regents of the University of Wisconsin System. *https://wida.wisc.edu/sites/default/files/resource/WIDA-ELD-Standards-Framework-2020.pdf.*

4

HABITATS LOCAL AND FAR AWAY LESSON PLANS

Andrea R. Milner, Vanessa B. Morrison, Janet B. Walton,
Carla C. Johnson, and Erin E. Peters-Burton

Lesson Plan 1:
Earth's Amazing Endangered Species

In this lesson, students will explore the basic needs of living things through the lens of habitats. Students will take a walking tour of the school neighborhood to identify local habitats and will connect the concepts of habitats and endangered species through an interactive game and through readings and videos. Students will be introduced to the module challenge.

ESSENTIAL QUESTIONS

- What is a species?

- What does endangered species mean?

- What different types of endangered species are there?

- Where and when have you seen endangered species?

- What are habitats?

- What kinds of habitats are there?

- Where and when have you seen various habitats?

- What are the basic needs of living things?

DOI: 10.4324/9781003450184-6

ESTABLISHED GOALS AND OBJECTIVES

At the conclusion of this lesson, students will be able to do the following:

- Explain what is meant by endangered species and identify several endangered species

- Identify several habitats and their characteristics

- Identify the basic needs of living things

- Observe and describe features of local habitats

- Describe how habitats provide living things with their basic needs

TIME REQUIRED

6 days (approximately 30 minutes each; see Tables 3.6–3.7, pp. 35–36)

MATERIALS

Required Materials for Lesson 1

- STEM Research Notebooks (1 per student, see pp. 12–13 for STEM Research Notebook information)

- Computer with internet access for viewing videos

- Books
  o *Nature's patchwork quilt: Understanding habitats* by Mary Miche
  o *Nature in the neighborhood* by Gordon Morrison
  o *Me and the measure of things* by Joan Sweeney
  o *Counting on Frank* by Rod Clement

- Chart paper

- Markers

- U.S. map and globe or world map

Additional Materials for Five Needs Game

- Index cards (50 for each team of 4 students)

Additional Materials for Neighborhood Explorers Activity

- Binoculars (1 for every 2 students)

- Magnifying glass (1 for every 2 students)

- Clipboard (for each student)

- Pencil (for each student)

SAFETY NOTES

1. Instruct students to be aware of and avoid poisonous plants and insects, any refuse, sharps (broken glass), and other hazards when they are outdoors.

2. Instruct students in safe use of binoculars and magnifying glasses, including instructing them not to look at the sun through the binoculars and not to concentrate the sun through the magnifying glass as this can create a great deal of heat and result in burns and even fire.

3. Have students wash hands with soap and water after activities are completed.

CONTENT STANDARDS AND KEY VOCABULARY

Table 4.1 lists the content standards from the *Next Generation Science Standards* (*NGSS*), *Common Core State Standards* (*CCSS*), the National Association for the Education of Young Children (NAEYC), and the Framework for 21st Century Learning that this lesson addresses, and Table 4.2 (p. 44) presents the key vocabulary. Vocabulary terms are provided for both teacher and student use. Teachers may choose to introduce some or all of the terms to students.

Table 4.1. Content Standards Addressed in STEM Road Map Module Lesson 1

NEXT GENERATION SCIENCE STANDARDS
PERFORMANCE OBJECTIVES
• LS1–2. Read texts and use media to determine patterns in behavior of parents and offspring that help offspring survive.
DISCIPLINARY CORE IDEAS
LS1.A. Structure and Function
• All organisms have external parts. Different animals use their body parts in different ways to see, hear, grasp objects, protect themselves, move from place to place, and seek, find, and take in food, water and air. Plants also have different parts (roots, stems, leaves, flowers, fruits) that help them survive and grow.
LS1.B. Growth and Development of Organisms
• Adult plants and animals can have young. In many kinds of animals, parents and the offspring themselves engage in behaviors that help the offspring to survive.
LS1.D. Information Processing
• Animals have body parts that capture and convey different kinds of information needed for growth and survival. Animals respond to these inputs with behaviors that help them survive. Plants also respond to some external inputs.

Continued

Table 4.1. (*continued*)

CROSSCUTTING CONCEPTS

Patterns
- Patterns in the natural and human designed world can be observed, used to describe phenomena, and used as evidence.

Structure and Function
- The shape and stability of structures of natural and designed objects are related to their function(s).

SCIENCE AND ENGINEERING PRACTICES

Constructing Explanations and Designing Solutions
- Constructing explanations and designing solutions in K–2 builds on prior experiences and progresses to the use of evidence and ideas in constructing evidence-based accounts of natural phenomena and designing solutions.
- Use materials to design a device that solves a specific problem or a solution to a specific problem. (1-LS1–1)

Obtaining, Evaluating, and Communicating Information
- Obtaining, evaluating, and communicating information in K–2 builds on prior experiences and uses observations and texts to communicate new information.
- Read grade-appropriate texts and use media to obtain scientific information to determine patterns in the natural world. (1-LS1–2)

COMMON CORE STATE STANDARDS FOR MATHEMATICS

MATHEMATICAL PRACTICES
- MP1. Make sense of problems and persevere in solving them.
- MP2. Reason abstractly and quantitatively.
- MP3. Construct viable arguments and critique the reasoning of others.
- MP4. Model with mathematics.
- MP5. Use appropriate tools strategically.
- MP6. Attend to precision.
- MP7. Look for and make use of structure.
- MP8. Look for and express regularity in repeated reasoning.

MATHEMATICAL CONTENT
- NBT.B.3. Compare two two-digit numbers based on meanings of the tens and ones digits, recording the results of comparisons with the symbols >, =, and <.
- NBT.C.5. Given a two-digit number, mentally find 10 more or 10 less than the number, without having to count; explain the reasoning used.
- NBT.C.6. Subtract multiples of 10 in the range 10–90 from multiples of 10 in the range 10–90 (positive or zero differences), using concrete models or drawings and strategies based on place value, properties of operations, and/or the relationship between addition and subtraction; relate the strategy to a written method and explain the reasoning used.
- 1.MD.C.4. Organize, represent, and interpret data with up to three categories; ask and answer questions about the total number of data points, how many in each category, and how many more or less are in one category than in another.

- OA.A.1. Use addition and subtraction within 20 to solve word problems involving situations of adding to, taking from, putting together, taking apart, and comparing, with unknowns in all positions.
- OA.A.2. Solve word problems that call for addition of three whole numbers whose sum is less than or equal to 20, e.g., by using objects, drawings, and equations with a symbol for the unknown number to represent the problem.

COMMON CORE STATE STANDARDS FOR ENGLISH LANGUAGE ARTS

READING STANDARDS
- RI.1.1. Ask and answer questions about key details in a text.
- RI.1.2. Identify the main topic and retell key details of a text.
- RI.1.3. Describe the connection between two individuals, events, ideas, or pieces of information in a text.
- RI.1.7. Use the illustrations and details in a text to describe its key ideas.

WRITING STANDARDS
- W.1.2. Write informative/explanatory texts in which they name a topic, supply some facts about the topic, and provide some sense of closure.
- W.1.6. With guidance and support from adults, use a variety of digital tools to produce and publish writing, including in collaboration with peers.
- W.1.7. Participate in shared research and writing.
- W.1.8. With guidance and support from adults, recall information from experiences or gather information from provided sources to answer a question.

SPEAKING AND LISTENING STANDARDS
- SL.1.1. Participate in collaborative conversations with diverse partners about grade 1 topics and texts with peers and adults in small and larger groups.
- SL.1.1.A. Follow agreed-upon rules for discussions.
- SL.1.1.B. Build on others' talk in conversations by responding to the comments of others through multiple exchanges.
- SL.1.1.C. Ask questions to clear up any confusion about the topics and texts under discussion.
- SL.1.3. Ask and answer questions about what a speaker says in order to gather additional information or clarify something that is not understood.
- SL.1.5. Add drawings or other visual displays to descriptions when appropriate to clarify ideas, thoughts, and feelings.

NATIONAL ASSOCIATION FOR THE EDUCATION OF YOUNG CHILDREN STANDARDS
- 2.E.1. Arrange firsthand, meaningful experiences that are intellectually and creatively stimulating, invite exploration and investigation, and engage children's active, sustained involvement by providing a rich variety of material, challenges, and ideas.
- 2.F.3. Extend the range of children's interests and the scope of their thought, present novel experiences and introduce stimulating ideas, problems, experiences, or hypotheses.

Continued

Table 4.1. (*continued*)

- 2.F.6. Enhance children's conceptual understanding through various strategies, including intensive interview and conversation, encourage children to reflect on and "revisit" their experiences.
- 2.G.2. Scaffolding takes on a variety of forms.
- 2.J.1. Incorporate a wide variety of experiences, materials and equipment, and teaching strategies to accommodate the range of children's individual differences in development, skills and abilities, prior experiences, needs, and interests.
- 3.A.1. Teachers consider what children should know, understand, and be able to do across the domains.

FRAMEWORK FOR 21ST CENTURY LEARNING
- Interdisciplinary Themes
- Learning and Innovation Skills
- Information, Media and Technology Skills
- Life and Career Skills

Table 4.2. Key Vocabulary in Lesson 1

Key Vocabulary	Definition
binoculars	a device that people can look through with both eyes that makes distant objects appear closer
climate	the weather conditions in an area over an extended period of time
continent	the continuous landmasses on Earth (Africa, Antarctica, Asia, Australia, Europe, North America, and South America)
endangered species	a plant or animal that is at risk of becoming extinct
extinct	describes an animal or plant when there is no longer any of that animal or plant left on Earth
habitat	a place in nature where plants and animals live and have their needs met
magnifying glass	a piece of glass that is specially shaped so that it makes an object held beneath it look larger
measure	determining the size or amount of something by comparing it with a known size or amount
species	a group of animals or plants that share the same characteristics
weather	the daily conditions over a particular area, including temperature, precipitation, cloud cover, and air pressure

TEACHER BACKGROUND INFORMATION

First graders are able to make connections across multiple content areas as well as the various developmental domains (physical, social and emotional, personality, cognitive, and language). Incorporating students' prior knowledge with developmentally appropriate instruction will enable them to make these connections. Through this lesson, you should support and facilitate the advancement of these domains within each student.

Giving students regular opportunities to share their ideas, discoveries, and questions about scientific phenomena in a group setting can build young learners' enthusiasm for topics, expand their perspectives, help them develop new ideas, and motivate them for scientific inquiry as well as helping them to build their language and communication skills. Asking open-ended questions about students' learning and ideas and encouraging them to make and share their observations are effective ways to engage kindergarten students in science talk, and allows opportunities for you to model using scientific language (terms associated with the concept being investigated) and practices (for example, observing, comparing, predicting, and measuring). In addition, this science talk can help you to identify misconceptions students may hold and provide opportunities for building new understanding. It may be helpful to work as a class to establish a set of guidelines or rules for this communication (for example, don't talk over another person who is speaking, be respectful of others' ideas even if we think they are wrong, look at the person talking, always ask questions when you have them).

Endangered Species

This module focuses on animals that are categorized as endangered species. The U.S. Fish and Wildlife Service provides a listing by state of animals in the U.S. that are endangered (see *www.fws.gov/endangered*). The World Wildlife Fund provides a global listing of endangered species at *www.worldwildlife.org/species/directory?direction=desc& sort=extinction_status*. For more information about endangered species and conservation, see the following websites:

- *www.earthsendangered.com/index.asp*

- *www.worldwildlife.org/species*

Careers

In this module, students will be introduced to the idea that engineers and other STEM workers work together in teams to solve problems. Students will experience working in teams and in pairs as they progress through a simple scientific process including predicting, observing, and explaining phenomena related to forces in this lesson. This introduction to teamwork sets the stage for students' use of the engineering design process (EDP) later in the module.

You may also wish to connect students' work in this module with other careers such as the following (adapted from Koehler, Bloom, and Milner 2015):

- Ecologist

- Geographer

- Journalist

- Mathematician

- Meteorologist

- Biologist

- Botanist

For more information about these and other careers, see the Bureau of Labor Statistics' *Occupational Outlook Handbook* at *www.bls.gov/ooh/home.htm*.

Know, Want to Know, Learned (KWL) Charts

Throughout this module, you will track student knowledge on Know, Want to Know, Learned (KWL) charts. These charts will be used to access and assess student prior knowledge, encourage students to think critically about the topic under discussion, and track student learning throughout the module. Each chart should consist of three columns, labeled "What We Know," "What We Want to Know," and "What We Learned." Write the topic at the top of each chart. It may be helpful to post these charts in a prominent place in the classroom so that students can refer to them throughout the module. Students will include their personal "know," "want to know," and "learned" reflections in their STEM Research Notebook entries. As students suggest content for the "want to know" portion of the charts, review lesson content to ensure that this content is covered and be prepared to provide additional content to address what students want to know.

Interactive Read Alouds

This module also uses interactive read alouds to engage students, access their prior knowledge, develop student background knowledge, and introduce topical vocabulary. These read alouds expose children to teacher-read literature that may be beyond their independent reading levels but is consistent with their listening level. Interactive read alouds may incorporate a variety of techniques, and you can find helpful information regarding these techniques at the following websites:

- *www.readingrockets.org/article/repeated-interactive-read-alouds-preschool-and-kindergarten*

- *www.readwritethink.org/professional-development/strategy-guides/teacher-read-aloud-that-30799.html*

4

In general, interactive read alouds provide opportunities for students to share prior knowledge and experiences, interact with the text and concepts introduced therein, launch conversations about the topics introduced, construct meaning, make predictions, and draw comparisons. You may wish to mark places within the texts to pause to ask for student experiences, predictions, or other ideas. Each reading experience should focus on an ongoing interaction between students and the text, including the following:

- Allow students to share personal stories throughout the reading.

- Ask students to predict throughout the story.

- Allow students to add new ideas from the book to the KWL chart and their STEM Research Notebooks.

- Allow students to add new words from the book to the vocabulary chart and their STEM Research Notebooks.

The materials list for each lesson includes the books for interactive read alouds that you will use in that lesson. A list of suggested books for additional reading can be found at the end of this chapter (see p. 109). You may wish to request books used for interactive read alouds and student research (listed in the Materials section of each lesson) from the school or public library in advance.

COMMON MISCONCEPTIONS

Students will have various types of prior knowledge about the concepts introduced in this lesson. Table 4.3 outlines common misconceptions students may have concerning these concepts. Because of the breadth of students' experiences, it is not possible to anticipate every misconception that students may bring as they approach this lesson. Incorrect or inaccurate prior understanding of concepts can influence student learning in the future, however, so it is important to be alert to misconceptions such as those presented in the table.

Table 4.3. Common Misconceptions About the Concepts in Lesson 1

Topic	Student Misconception	Explanation
Habitats	An animal's home is its habitat.	An animal's home is part of its habitat and provides shelter but does not provide all of its basic needs. The surrounding environment, which includes the animal's home, is its habitat.
Animals	Insects are not animals.	Insects are animals, and in fact over half of all living things are insects.

PREPARATION FOR LESSON 1

Review the Teacher Background Information, assemble the materials for the lesson, duplicate the student handouts, and preview the videos recommended in the Learning Plan Components section below. Present students with their STEM Research Notebooks and explain how these will be used (see pp. 12–13). Templates for the STEM Research Notebook are provided in Appendix A, and a rubric for observations, student participation, and STEM Research Notebook entries is provided in Appendix B. Throughout the module, be prepared to read aloud STEM Research Notebook instructions and provide support for students as they create their entries.

You will need to prepare index cards for the Five Needs game (see Activity/Investigation, p. 52). For each team of four students, prepare 10 index cards labeled "food," 10 index cards labeled "water," 10 index cards labeled "shelter/habitat," 10 index cards labeled "air," and 10 index cards labeled "sunlight." These cards should be shuffled together.

STEM Research Notebook Entry #3 provides a template for students to record vocabulary words. You may wish to use this template throughout the module for students to record definitions and illustrations of key vocabulary words as they are introduced. Alternatively, you may wish to make a class chart of vocabulary words and choose one or two of the words on the chart for each lesson and have students write and draw each word on a separate piece of paper to keep in their notebooks. The template provides space for definitions and illustrations of three words. If you introduce more than three vocabulary words in a lesson, you should make multiple copies of the template for each student.

Be prepared to discuss endangered species with students (see Teacher Background Information, pp. 45–57). Have on hand examples of endangered animal species such as those included in the video the students view during the module launch (see Introductory Activity/Engagement, p. 49) and local animal species that are endangered.

Students will be outdoors for the Neighborhood Explorers activity (see Activity/Investigation, pp. 53–54). You should familiarize yourself with your school's policy on outdoors activities, and engage adult volunteers for additional student support and supervision. If your school policies prohibit you from leaving the school grounds, an option is to make a videorecorded tour of the neighborhood that the class can view together and limit students' firsthand exploration to the school property. In addition, you should check the weather forecast and make appropriate preparations (for example, having students wear hats, carry umbrellas).

LEARNING PLAN COMPONENTS
Introductory Activity/Engagement

Connection to the Challenge: After introducing the module challenge, begin each day of this lesson by directing students' attention to the module challenge, the Save the Species Challenge:

There are hundreds of endangered species all over the world. If people don't do something, some of these species may become extinct. Your challenge is to develop an action plan about what people can do to help this species survive in its habitat.

At the start of the lesson, ask students what they know and what they wonder about endangered species, creating a KWL chart. Next, ask students for their ideas about what they will need to know to create their action plans, creating a class list of students' ideas and adding these to the "want to know" section of the KWL chart. Each day thereafter, hold a brief class discussion of how students' learning in the previous days' lessons contributed to their ability to complete the challenge, and add students' responses to the "learned" section of the KWL chart.

Science and Social Studies Classes: Introduce the module by holding a class discussion about habitats. Create a Know/Want to Learn/Learned (KWL) chart to record students' ideas. Following agreed-upon rules for discussions, launch the module by holding a class discussion about endangered species, asking students these questions:

- What is a species?
- What do we mean by endangered species?
- Are there different types of endangered species?
- Why might an animal species be endangered?
- Can a plant be an endangered species?
- Where and when have you seen endangered species?

Tell students that they will watch a video about animals that are endangered species. Before showing the video, ask students to predict what type of animals will be on the video. Track students' responses on a KWL chart. Next, show a video depicting various endangered species such as "Top 10 Most Endangered Species" at *www.youtube. com/watch?v=fm8qTACshos*. After watching the video, compare students' predictions with the animals they saw in the video.

Next, have each student complete STEM Research Notebook Entry #1.

STEM Research Notebook Entry #1

Have students draw and label one endangered animal, and write one thing that they know about this animal in their STEM Research Notebooks.

Next, have students share the animal they drew with the class, sharing one thing they know about the species and one thing they wonder. Document student responses on a KWL chart.

Ask students where they think endangered species live. Guide students to understand that endangered species are everywhere. Tell students that there may be species that are endangered right in the local community (you may wish to provide photos of several local endangered species; for example, the Indiana bat) and all over the world.

Next, hold a class discussion about habitats, asking students to respond to the following questions and recording their responses on a KWL chart:

- What are habitats?

- Are there different types of habitats?

- What different kinds of habitats are there?

- Where and when have you seen various habitats?

Using a globe or world map, help students identify the seven continents (Africa, Antarctica, Asia, Australia, Europe, North America, and South America). Ask students to identify habitats that are associated with the different continents (e.g., deserts, rainforests, oceans, forests, tundra). Ask students for their ideas about what endangered species may live in those habitats. Discuss why these continents are ideal locations for those specific habitats and animals (e.g., climate).

Next, students will explore habitats through an interactive read aloud of *Nature's Patchwork Quilt: Understanding Habitats* by Mary Miche. After the read aloud, ask students to share what they learned about habitats, recording students' responses on the KWL chart and having students complete a STEM Research Notebook entry.

STEM Research Notebook Entry #2

Have students record what they learned about habitats in their STEM Research Notebooks after the interactive read aloud, using both words and pictures.

Mathematics Connection: Ask students for their ideas about how we can describe habitats, using a desert habitat as an example. Record students' responses on a class chart. Students will likely respond with words such as "dry," "hot," and "sandy." Next, ask students for their ideas about how a desert habitat could be described using numbers in addition to words, focusing students' attention on what kinds of things in habitats can be counted or measured (e.g., temperature, number of rainy days each year, how many different kinds of animals and plants live there), adding students' responses to the chart.

ELA Connection: Through group discussion, students will utilize and develop speaking and listening skills throughout the module. In addition, students will utilize and develop their reading and writing skills through their STEM Research Notebook entries and will build vocabulary knowledge through class discussions and readings.

Begin a class vocabulary chart using pictures and words. Add vocabulary words to this chart throughout the module and refer to the chart during class discussions and as students create their STEM Research Notebook entries. This chart should be posted on the classroom wall throughout the module.

STEM Research Notebook Entry #3

(see Preparation for Lesson 1, p. 48 for options for vocabulary words)

Have students record vocabulary terms and definitions using both words and pictures.

Activity/Investigation

Science and Social Studies Classes and ELA Connection: Revisit the discussion about continents in the Introductory Activity/Engagement and discuss the cultural and societal implications of living in different places. For example, ask students:

- In what types of habitats would we find more people living? Why or why not? Where are these habitats located?

- How would life be the same and different in various habitats?

- What are the pros and cons of living in various habitats?

Record students' responses on chart paper.

Ask students for their ideas about why different types of animals live in different habitats. Guide students to understand that different habitats have different conditions. Ask students for their ideas about what these different conditions might be (e.g., climate, amount of water, types of plants, mountains, or flat land). Introduce the idea to students that all living things have basic needs and that animals live in habitats that can meet these needs. Ask students for their ideas about the basic needs of living things, creating a class list. The class list of living things should include:

- Food

- Water

- Air

- Shelter

- Space

As a class, choose an animal that lives in your local area (e.g., squirrel, bird) and review how the local habitat meets each of that animal's needs. Next, have students create a STEM Research Notebook entry in which they record the five basic needs.

STEM Research Notebook Entry #4

Have students document their learning about how each of the five basic needs of an animal are met in its habitat, using both words and pictures.

Five Needs Game

Introduce the Five Needs game by reviewing the basic needs of living things and posting the list of the five basic needs where students can see it. Next, group students in teams of four at tables. Distribute the index cards you prepared (see Preparation for Lesson 1, p. 48) to each team. Teams should each have 10 index cards labeled "food," 10 index cards labeled "water," 10 index cards labeled "shelter/ habitat," 10 index cards labeled "air," and 10 index cards labeled "sunlight." These cards should be shuffled together. Tell students that they are going to play a card game, and the goal is to have all of their basic needs met. Guide students through the following procedure:

- Have all students stand up.

- Have one student from each team pull eight cards from the deck.

- If that student has cards that read "food," "water," "shelter/habitat," "air" *and* "sunlight" they can remain standing. If they do not have a card with each label, they must sit (their basic needs have not been met so they are now endangered).

- Put the cards that have been pulled into a separate pile *not* to be used again.

- Have the second person from each table repeat.

- Continue until there is only one person left standing at each table (when the card pile is depleted, have students reshuffle the cards to renew the draw pile).

After students have completed several rounds of the activity, ask students:

- What happens when all five basic needs of living things are not met?

- Why is it important for all living things to have these basic needs met?

- Why do you think some living things become extinct?

Document student responses on chart paper. Students will also record their ideas about the importance of the five basic needs in a STEM Research Notebook Entry.

STEM Research Notebook Entry #5

Have students record the five basic needs of living things and the effects on animals when these needs are not met, using both words and pictures.

Mathematics Connection: Continue the discussion you began in the Introductory Activity/Engagement about how habitats can be described using numbers. Conduct an interactive read aloud of the book *Me and the Measure of Things* by Joan Sweeney. Then, as a class, work to describe the classroom using numbers (e.g., number of windows, number of tables, size of classroom, number of students).

Explanation

Science and Social Studies Classes: Discuss the Five Needs game, asking students questions such as:

- Were every student's basic needs met all the time?
- Why do you think that not all basic needs could be met for each student?
- What are some reasons animals' basic needs may not be met?
- What can people do to make sure the basic needs of living things are met?

Have students share their ideas and explanations with the whole class, documenting student responses on chart paper.

Neighborhood Explorers Activity

Introduce the Neighborhood Explorers activity by conducting an interactive read aloud of *Nature in the Neighborhood* by Gordon Morrison. After the read aloud, ask students to describe the living things and habitat they observed in the book. Tell students that in this activity they will explore the schoolyard or local neighborhood to identify animals and features of their habitats (e.g., insect habitats, bird habitats, squirrel habitats, earthworm habitats).

Ask students for their ideas about what tools they might need to observe animals and habitats. Introduce the idea that scientists use tools to observe things and take measurements, and that special tools are necessary to observe things that are far away or that are very small. Ask students for their ideas about what tools scientists use to observe things that are far away or very small (e.g., telescope, microscope). Tell students that the tools they will use for this activity are binoculars and magnifying glasses. Demonstrate using each of these and ask students for their ideas about why students might need these to see animals and habitats. Demonstrate the proper use of each and review safety guidelines (don't look at the sun through the binoculars, don't concentrate the sun through the magnifying glass as this can create a great deal of heat and even fire).

Distribute binoculars and magnifying glasses to each student and have them practice using these in the classroom. Next, distribute clipboards to each student along with STEM Research Notebook Entry #6. Have students take these, along with a

pencil for each student and the binoculars and magnifying glasses for each pair, and take students outside. Students should work in pairs to identify at least two animals and their habitats. Students will draw pictures of the animals and habitats they observed and label the pictures with how the animals' five basic needs are met in the habitats.

STEM Research Notebook Entry #6

Have students each draw pictures of the animals and habitats they observed and label the pictures with how the animals' five basic needs are met in the habitats.

Mathematics Connection: Hold a class discussion about how students could describe their observations in the Neighborhood Explorers activity numerically. Next, conduct an interactive read aloud of *Counting on Frank* by Rod Clement. After the read aloud, ask students for their ideas about how the character in the book answered questions using mathematics. Next, ask students if they have any new ideas about how they could answer the question, "What is our neighborhood habitat like?" using mathematics and measurement. Ask students for their ideas about what tools they could to measure their observations, creating a class list (for example, ruler, scale, thermometer).

Elaboration/Application of Knowledge

Science and Social Studies Classes and ELA Connection: Ask students to use their understanding of the basic needs of living things to answer the question, "Why do some animals become endangered?" Create a class list of students' ideas on a KWL chart. Ask students what they wonder about why animals become endangered, again recording students' ideas on the KWL chart.

Next, have students explore reasons that animals become endangered by watching a video such as "Why animals become endangered" at *www.youtube.com/watch?v=So62I2dJZyo*. After watching the video, ask students to share what they learned about why animals become endangered, documenting student responses on a KWL chart and having students document their learning in a STEM Research Notebook entry.

STEM Research Notebook Entry #7

Have students document their learning about why animals become endangered in their STEM Research Notebooks, using both words and pictures.

Assess student learning by having students draw an animal in its habitat and label how its basic needs are met in this habitat.

Mathematics Connection: Ask students how endangered animals could be described using mathematics (for example, the number of animals left, how much of their habitat has been destroyed). Again show the video about reasons for animal endangerment, this time asking students to pay attention to how endangered animals are described using numbers. Hold a class discussion about mathematical descriptions students noticed in the video.

Evaluation/Assessment

Students may be assessed on the following performance tasks and other measures listed.

Performance Tasks

- Neighborhood Explorers activity

- Lesson assessment

Other Measures (using assessment rubric in Appendix B)

- Teacher observations

- STEM Research Notebook entries

- Participation in teams during investigations

INTERNET RESOURCES

Endangered animals lists

- *www.fws.gov/endangered/*

- *www.worldwildlife.org/species/directory?direction=desc&sort=extinction_status*

Endangered species information

- *www.earthsendangered.com/index.asp*

- *www.worldwildlife.org/species*

Bureau of Labor Statistics' *Occupational Outlook Handbook*

- *www.bls.gov/ooh/home.htm*

Interactive read alouds

- *www.readingrockets.org/article/repeated-interactive-read-alouds-preschool-and-kindergarten*

- *www.readwritethink.org/professional-development/strategy-guides/teacher-read-aloud-that-30799.html*

"Top 10 most endangered species" video

- *www.youtube.com/watch?v=fm8qTACshos*

"Why animals become endangered" video

- *www.youtube.com/watch?v=So62I2dJZyo*

4

Lesson Plan 2:
Our Local Endangered Species

In this lesson, students will explore endangered species at the local level. Students will use the engineering design process (EDP) to create a model of a habitat for one of these species using recycled materials.

ESSENTIAL QUESTIONS

- What animal species are endangered where we live?

- What causes species to become endangered?

- How can we describe the habitat where we live?

- What is the weather and climate like where we live?

- What types of animals and plants live where we live?

- What time of year is it? How can you tell?

ESTABLISHED GOALS AND OBJECTIVES

At the conclusion of this lesson, students will be able to do the following:

- Identify local endangered animal species

- Identify several reasons for animal species becoming endangered

- Identify ways that humans can help to prevent endangered species from becoming extinct

- Design and construct mobiles depicting local endangered animal species

- Use the steps of the EDP to create a model habitat for a local endangered animal species

- Identify engineering as a career and describe the kind of work engineers do

- Identify tools that scientists use to investigate the natural world

- Identify several examples of interdependency in ecosystems

- Depict a simple food chain using words and drawings

TIME REQUIRED

7 days (approximately 30 minutes each; see Tables 3.7–3.8, p. 36)

MATERIALS

Required Materials for Lesson 2

- STEM Research Notebooks

- Computer with internet access for viewing videos

- Books

  o *The buffalo are back* by Jean Craighead George
  o *Rosie Revere, engineer* by Andrea Beaty

- Chart paper

- Markers

- U.S. map and world map or globe

- Bat box

Additional Materials for Mighty Mobiles Activity (for each team of 3–4 students)

- 1 paper plate

- 3–4 lengths of yarn or string of varying lengths

- 3–4 glue sticks

- 1 roll of masking tape

- 3–4 half sheets of construction paper

- 3–4 pairs of scissors

Additional Materials for Habitat Helpers Activity (for each team of 3–4 students)

- Access to a class set of recycled materials

- 3–4 glue sticks

- 5 sheets of construction paper

- 10 chenille stems

- 3–4 pairs of scissors

- Materials for habitats specific to your area (e.g., sand, small stones)

- 2 11 x 13 sheets of foam board, attached at 90 degree angles to create a surface and backdrop on which to construct the habitat

SAFETY NOTES

1. Students should use caution when handling scissors, as the sharp points and blades can cut or puncture skin.

2. Tell students to be careful when handling recycled bottles and cans. Cans may have sharp edges, which can cut or puncture skin. Glass or plastic bottles can break and cut skin.

3. Have students wash hands with soap and water after the activity is completed.

CONTENT STANDARDS AND KEY VOCABULARY

Table 4.4 lists the content standards from the *NGSS*, *CCSS*, NAEYC, and the Framework for 21st Century Learning that this lesson addresses, and Table 4.5 presents the key vocabulary.

Vocabulary terms are provided for both teacher and student use. Teachers may choose to introduce some or all of the terms to students.

Table 4.4. Content Standards Addressed in STEM Road Map Module Lesson 2

NEXT GENERATION SCIENCE STANDARDS

PERFORMANCE OBJECTIVES
- LS1–2. Read texts and use media to determine patterns in behavior of parents and offspring that help offspring survive.
- K-2 ETS1–2. Develop a simple sketch, drawing, or physical model to illustrate how the shape of an object helps it function as needed to solve a given problem.

DISCIPLINARY CORE IDEAS

LS1.A. Structure and Function
- All organisms have external parts. Different animals use their body parts in different ways to see, hear, grasp objects, protect themselves, move from place to place, and seek, find, and take in food, water and air. Plants also have different parts (roots, stems, leaves, flowers, fruits) that help them survive and grow.

LS1.B. Growth and Development of Organisms
- Adult plants and animals can have young. In many kinds of animals, parents and the offspring themselves engage in behaviors that help the offspring to survive.

LS1.D. Information Processing
- Animals have body parts that capture and convey different kinds of information needed for growth and survival. Animals respond to these inputs with behaviors that help them survive. Plants also respond to some external inputs.

Continued

Table 4.4. (*continued*)

CROSSCUTTING CONCEPTS

Patterns
- Patterns in the natural and human designed world can be observed, used to describe phenomena, and used as evidence.

Structure and Function
- The shape and stability of structures of natural and designed objects are related to their function(s).

SCIENCE AND ENGINEERING PRACTICES

Constructing Explanations and Designing Solutions
- Constructing explanations and designing solutions in K–2 builds on prior experiences and progresses to the use of evidence and ideas in constructing evidence-based accounts of natural phenomena and designing solutions.
- Use materials to design a device that solves a specific problem or a solution to a specific problem.

Obtaining, Evaluating, and Communicating Information
- Obtaining, evaluating, and communicating information in K–2 builds on prior experiences and uses observations and texts to communicate new information.
- Read grade-appropriate texts and use media to obtain scientific information to determine patterns in the natural world.

Developing and Using Models
- Modeling in K–2 builds on prior experiences and progresses to include using and developing models (i.e., diagram, drawing, physical replica, diorama, dramatization, storyboard) that represent concrete events or design solutions.
- Use a model to represent relationships in the natural world.

COMMON CORE STATE STANDARDS FOR MATHEMATICS

MATHEMATICAL PRACTICES
- MP1. Make sense of problems and persevere in solving them.
- MP2. Reason abstractly and quantitatively.
- MP3. Construct viable arguments and critique the reasoning of others.
- MP4. Model with mathematics.
- MP5. Use appropriate tools strategically.
- MP6. Attend to precision.
- MP7. Look for and make use of structure.
- MP8. Look for and express regularity in repeated reasoning.

MATHEMATICAL CONTENT
- NBT.B.3. Compare two two-digit numbers based on meanings of the tens and ones digits, recording the results of comparisons with the symbols >, =, and <.
- NBT.C.5. Given a two-digit number, mentally find 10 more or 10 less than the number, without having to count; explain the reasoning used.

- NBT.C.6. Subtract multiples of 10 in the range 10–90 from multiples of 10 in the range 10–90 (positive or zero differences), using concrete models or drawings and strategies based on place value, properties of operations, and/or the relationship between addition and subtraction; relate the strategy to a written method and explain the reasoning used.
- 1.MD.C.4. Organize, represent, and interpret data with up to three categories; ask and answer questions about the total number of data points, how many in each category, and how many more or less are in one category than in another.
- OA.A.1. Use addition and subtraction within 20 to solve word problems involving situations of adding to, taking from, putting together, taking apart, and comparing, with unknowns in all positions.
- OA.A.2. Solve word problems that call for addition of three whole numbers whose sum is less than or equal to 20, e.g., by using objects, drawings, and equations with a symbol for the unknown number to represent the problem.

COMMON CORE STATE STANDARDS FOR ENGLISH LANGUAGE ARTS

READING STANDARDS
- RI.1.1. Ask and answer questions about key details in a text.
- RI.1.2. Identify the main topic and retell key details of a text.
- RI.1.3. Describe the connection between two individuals, events, ideas, or pieces of information in a text.
- RI.1.7. Use the illustrations and details in a text to describe its key ideas.

WRITING STANDARDS
- W.1.2. Write informative/explanatory texts in which they name a topic, supply some facts about the topic, and provide some sense of closure.
- W.1.6. With guidance and support from adults, use a variety of digital tools to produce and publish writing, including in collaboration with peers.
- W.1.7. Participate in shared research and writing.
- W.1.8. With guidance and support from adults, recall information from experiences or gather information from provided sources to answer a question.

SPEAKING AND LISTENING STANDARDS
- SL.1.1. Participate in collaborative conversations with diverse partners aboutgrade 1 topics and textswith peers and adults in small and larger groups.
- SL.1.1.A. Follow agreed-upon rules for discussions.
- SL.1.1.B. Build on others' talk in conversations by responding to the comments of others through multiple exchanges.
- SL.1.1.C. Ask questions to clear up any confusion about the topics and texts under discussion.
- SL.1.3. Ask and answer questions about what a speaker says in order to gather additional information or clarify something that is not understood.
- SL.1.5. Add drawings or other visual displays to descriptions when appropriate to clarify ideas, thoughts, and feelings.

Continued

Table 4.4. (*continued*)

> **NATIONAL ASSOCIATION FOR THE EDUCATION OF YOUNG CHILDREN STANDARDS**
> * 2.E.1. Arrange firsthand, meaningful experiences that are intellectually and creatively stimulating, invite exploration and investigation, and engage children's active, sustained involvement by providing a rich variety of material, challenges, and ideas.
> * 2.F.3. Extend the range of children's interests and the scope of their thought, present novel experiences and introduce stimulating ideas, problems, experiences, or hypotheses.
> * 2.F.6. Enhance children's conceptual understanding through various strategies, including intensive interview and conversation, encourage children to reflect on and "revisit" their experiences.
> * 2.G.2. Scaffolding takes on a variety of forms.
> * 2.J.1. Incorporate a wide variety of experiences, materials and equipment, and teaching strategies to accommodate the range of children's individual differences in development, skills and abilities, prior experiences, needs, and interests.
> * 3.A.1. Teachers consider what children should know, understand, and be able to do across the domains.
>
> **FRAMEWORK FOR 21ST CENTURY LEARNING**
> * Interdisciplinary Themes
> * Learning and Innovation Skills
> * Information, Media and Technology Skills
> * Life and Career Skills

Table 4.5. Key Vocabulary in Lesson 2

Key Vocabulary	Definition
conservation	working to protect or restore something, such as plants or animals, in the natural environment
ecosystem	the living and nonliving things in an area that interact with each other
interdependence	describes how things depend on each other to survive and be healthy
model	a simpler way to describe or show something that is complicated

TEACHER BACKGROUND INFORMATION
Endangered Species

The focus of this lesson will be on endangered species in your local area. According to the U.S. Environmental Protection Agency (EPA), there were about 1,300 endangered or threatened animal and plant species in the U.S. in 2022. Species can become endangered either by natural causes, such as volcanic eruptions or climate changes over time, or by human actions such as over-hunting, pollution, and habitat destruction. More information about endangered and threatened species is available on the EPA Endangered Species webpage at *www.epa.gov/endangered-species/learn-more-about-threatened-and-endangered-species*. The U.S. Fish and Wildlife Endangered Species webpage provides a list of endangered species by state at *www.fws.gov/endangered.*

The American bison is the focus of introductory activities for this lesson. The bison was declared the national mammal of the United States in 2016 and is an example of a successful conservation effort. The American bison is no longer an endangered species due to conservation efforts, and the population's status is considered stable although efforts continue to grow the population. The U.S. Department of the Interior provides information about the American bison and the population's resurgence from near extinction at *www.doi.gov/blog/15-facts-about-our-national-mammal-american-bison.*

Engineering

Students begin to gain an understanding of engineering as a profession in this lesson as they learn to use the engineering design process (EDP) to plan and construct model habitats for a local endangered species in the Habitat Helpers activity. Students should understand that engineers are people who design and build products and systems in response to human needs. For an overview of the various types of engineering professions, see the following websites:

- *www.engineergirl.org/33/TryOnACareer*

- *www.sciencekids.co.nz/sciencefacts/engineering/typesofengineeringjobs.html*

Engineering Design Process (EDP)

Students should understand that engineers need to work in groups to accomplish their work, and that collaboration is important for designing solutions to problems. In this lesson and the next one, students will use the engineering design process (EDP), the same process that professional engineers use in their work. A graphic representation of the EDP is provided at the end of this lesson. You may wish to provide each student with a copy of the EDP graphic or enlarge it and post it in a prominent place in your classroom for student reference throughout the module. Be prepared to review each step of the EDP listed on the graphic with students and emphasize that the process is not a linear one – at any point in the process, they may need to return to a previous step. The steps of the process are as follows:

1. *Define.* Describe the problem you are trying to solve, identify what materials you are able to use, and decide how much time and help you have to solve the problem.

2. *Learn.* Brainstorm solutions and conduct research to learn about the problem you are trying to solve.

3. *Plan.* Plan your work, including making sketches and dividing tasks among team members if necessary.

4. *Try.* Build a device, create a system, or complete a product.

5. *Test.* Now, test your solution. This might be done by conducting a performance test, if you have created a device to accomplish a task, or by asking for feedback from others about their solutions to the same problem.

6. *Decide.* Based on what you found out during the Test phase, you can adjust your solution or make changes to your device.

After completing all six steps, students share their solutions or devices with others. This represents an additional opportunity to receive feedback and make additional modifications based on that feedback.

The following are additional resources about the EDP:

- *www.sciencebuddies.org/engineering-design-process/engineering-design-compare-scientific-method.shtml*

- *www.pbslearningmedia.org/resource/phy03.sci.engin.design.desprocess/what-is-the-design-process*

COMMON MISCONCEPTIONS

Students will have various types of prior knowledge about the concepts introduced in this lesson. Table 4.6 outlines a common misconception students may have concerning these concepts. Because of the breadth of students' experiences, it is not possible to anticipate every misconception that students may bring as they approach this lesson.

Incorrect or inaccurate prior understanding of concepts can influence student learning in the future, however, so it is important to be alert to misconceptions such as that presented in the table.

Table 4.6. Common Misconception About the Concepts in Lesson 2

Topic	Student Misconception	Explanation
Engineers and the engineering design process (EDP)	All engineers are people who drive trains.	Railroad engineers are just one type of engineer. The engineers referred to in this module are people who use science, technology, and mathematics to build machines, products, and structures that meet people's needs.

PREPARATION FOR LESSON 2

Review the Teacher Background Information provided, assemble the materials for the lesson, and duplicate the EDP graphic attached at the end of this lesson plan (Figure 4.1) if you wish to hand this out to students or enlarge it to post in the classroom.

Be prepared with a list of local or regional endangered species (see *https://ecos. fws.gov/ecp/report/species-listings-by-state-totals?statusCategory=Listed*). Be prepared to assign each team of students an animal family and have a list of three to four animals in that animal family (one for each team member). Each team member will investigate an animal from that family (for example, birds, insects, mammals). Print pictures and information about each species to have available in the classroom. You may wish to engage adult volunteers for additional support as students conduct research in their teams.

Since many species are endangered because of habitat destruction, students will be challenged to design a model of a substitute habitat (similar to a simulated habitat found in a zoo) that meets an endangered animal's needs. Students will use recycled materials for this task. You should collect recyclable materials in advance; you may wish to have students bring in various materials from home (for example, boxes, cans, milk jugs, plastic bottles, packing materials, newspapers, fabric scraps).

In this lesson, you will show students a bat box as an example of a simulated habitat for an endangered species (the Indiana bat). Be prepared to introduce the discussion of bats with a picture of a bat (see *https://animalstime.com/facts-about-bats-kids-bats-diet-and-habitat*, for example). Acquire the bat box in advance and assemble it if necessary. Hardware stores and farm stores typically sell bat boxes. An option for this lesson is a field trip to hang the bat box or, alternatively, you may choose to donate the box to a local nature center. If you choose to hang the bat box, obtain approval for a location to hang the box and make appropriate preparations to install the box. To be effective at attracting bats, the bat boxes should be hung according to the following specifications:

- At least 10 feet off the ground

- In a south- or southeast-facing orientation

- Where they receive at least seven hours of sun a day

- Preferably within 1,500 feet of a stream or pond

- Within 10–30 yards of a tree line

Other options for this lesson include inviting a guest speaker such as a representative from a local nature center or a zoologist and/or taking a field trip related to an animal conservation-related site (see Elaboration/Application of Knowledge, p. 70). Make appropriate preparations for speakers and/or field trips.

LEARNING PLAN COMPONENTS
Introductory Activity/Engagement

Connection to the Challenge: Begin each day of this lesson by directing students' attention to the module challenge, the Save the Species Challenge:

> There are hundreds of endangered species all over the world. If people don't do something, some of these species may become extinct. Your challenge is to develop an action plan about what people can do to help this species survive in its habitat.

Hold a brief class discussion of how students' learning in the previous days' lessons contributed to their ability to complete the challenge, and add students' responses to the "learned" section of the KWL chart.

Science and Social Studies Classes and ELA Connection: Ask students if they have heard any news or advertisements about ways to help endangered species (for example, a save the whales campaign). Introduce the idea of conservation as human efforts to preserve natural habitats and help species survive.

Introduce the American bison as an example of conservation success. Tell students that the bison was once an endangered species but has now been taken off the endangered species list because the numbers of bison have increased due to conservation efforts. Students will explore conservation efforts related to the American bison through watching a video and an interactive read aloud.

First, show students an image of an American bison (attached at the end of this lesson plan). Ask students to share what they know about this animal, recording student responses on a KWL chart. Next, show a video about American bison such as "Meet the American bison" at *www.youtube.com/watch?v=2iVO5mkIn1c*. After watching the video, ask students what they learned about the bison, adding responses to the KWL chart.

Guide students in identifying where they live on a U.S. map and the region where most bison live (the Great Plains region, encompassing Montana, North Dakota, South Dakota, Wyoming, Nebraska, Kansas, Colorado, Oklahoma, Texas, and New Mexico). As a class, compare and contrast the habitat in your region with the region where bison live, creating a class list of similarities and differences. Have students share their ideas about why the American bison population declined in the past and why it has increased in recent years (add to the KWL chart).

Next, conduct an interactive read aloud of *The Buffalo are Back* by Jean Craighead George. After the read aloud, hold a class discussion about what students learned, adding to the KWL chart and having students complete a STEM Research Notebook entry.

<u>STEM Research Notebook Entry #8</u>

Have students document what they learned about the American bison in their STEM Research Notebooks, using both words and pictures.

Mathematics Connection: Review the video and interactive read aloud and ask students to point out how the American bison population is described using numbers. Create a class list of how mathematics is used to describe the bison and its habitat.

Activity/Investigation

Science and Social Studies Classes and ELA Connection: Students will investigate habitats by participating in two activities, Mighty Mobiles and Habitat Helpers. These activities focus on local endangered species and their habitats.

Introduce the activities by having students brainstorm how people, plants, and animals live in the local habitat (for example, ask "What is easy about living in this habitat? What might be difficult?"). Focus students' attention on the idea that there are endangered species in their local area.

Prepare for the Mighty Mobiles activity by asking students:

- Where do we live?

- What is the climate like where we live?

- What types of animals and plants live where we live?

- Are there any animals or plants that are endangered where we live?

Introduce the concepts of adaptations to students. Students should understand that adaptations are physical features or behaviors of animals that help them to survive in their environments. Focus students' attention on body parts, body coverings, and behaviors of animals. Have the class brainstorm ideas for each of these categories. For example, students might identify webbed feet as an adaptation for ducks, thick fur as an adaptation for bears that live in cold climates, and winter migration to warm climates for birds.

Mighty Mobiles

For the Mighty Mobiles activity, group students in teams of three to four. Students will work in these teams to create mobiles featuring local endangered species. Assign each team an animal family (see Teacher Background, pp. 62–64 and Preparation for Lesson 2, p. 65). Have students use the printed information about and pictures of local endangered species you compiled to complete STEM Notebook Entry #9. Students will use this information to create their mobiles.

STEM Research Notebook Entry #9

Have students use the printed information about and pictures of their local endangered species, recording the following information (using words and pictures) in their notebooks:

- Name of the endangered species

- The region or state where their endangered species lives

- Description of the habitat of the endangered species

- Reasons why this species is endangered

- Description of what people can do to ensure the survival of the endangered species

To create the mobile, each student should attach a picture (drawn or printed) of the endangered species to a half sheet of construction paper. On the back of the paper each student should include the following information:

- The name of the animal

- One reason the animal is endangered

- One thing people can do to help the animal species to survive

The team should decorate a paper plate and identify the animal family that their species belongs to on the plate.

Each student should then attach the picture of their endangered species to one end of a string, and then attach the strings to the paper plate to create their Mighty Mobile.

Have student teams present their Mighty Mobiles, introducing the animal family their team was assigned and have each team member present information about their endangered species.

Habitat Helpers Activity

Introduce the Habitat Helpers activity by asking students to name some of the reasons that their local endangered species are threatened. Point out that habitat change or destruction is a very common reason that species' survival is threatened. Tell students that they are going to work as teams to create a simulated habitat for one of the local endangered animals their team investigated.

Ask students to share their ideas about where they have seen simulated habitats. Point out to students that zoos create simulated habitats for animals. Other simulated habitats include bat boxes, backyard beehives, and goldfish ponds. These are habitats that humans create to ensure the survival of animals.

Share a picture of a bat with students (see Preparation for Lesson 2, p. 65). Ask students to share their ideas about bat habitats. Guide students to understand that bats often live in enclosed places that are protected from direct sunlight during the day such as caves. Have students share ideas about why cave habitats could be threatened (for example, construction, human presence in caves, pollution). Show students a bat box and ask them to identify features of the box that simulate a cave habitat. Point out to students that these boxes need to be placed in specific locations to ensure the survival of the bats. For instance, bat boxes must be hung at least ten feet off the ground so that ground animals don't interfere with the boxes. They should also be near water and near a wooded area (see Teacher Background Information, pp. 62–64 for more information).

Next, tell students that they will create a simulated habitat for one of the animals their team included in their Mighty Mobiles. Have students work in their Mighty Mobiles teams to choose one animal from their mobile for which they will create the simulated habitat.

Next, introduce students to the idea that they will act as engineers who are solving a problem (creating a habitat for an endangered species), and that they will follow a specific set of steps as they solve the problem. Introduce the EDP using the EDP graphic attached at the end of this lesson plan and the information provided in the Teacher Background Information section. Show students the recycled materials available for the activity (see Preparation for Lesson 2, pp. 62–64). Tell students that they will also have glue sticks and masking tape to use to create their simulated habitats.

Ask students for their ideas about whether they will be able to create a full-sized habitat for their species or a habitat that animals could actually live in. Introduce the idea of modeling, telling students that the habitats they create will be models that will include important details about the habitat but that may not be full sized and that are not intended for animals to actually live in. Ask students for their ideas about where they have seen models used (for example, a globe or map). Hold a class discussion about the advantages of using models (models can represent things that are difficult to see or understand in ways that are easier to see and understand). Next, ask students to share their ideas about disadvantages of models (for example, they don't show all the parts of the real object or system).

Guide student teams through each step of the EDP. Students will record their progress in a STEM Research Notebook entry.

STEM Research Notebook Entry #10

Have students record their progress through the steps of the EDP (define, learn, plan, try, test, decide, and share) as they create a simulated habitat for a local endangered species in the Habitat Helpers activity.

After student teams have created their habitat models, have each team share the habitat they created with the class, identifying the endangered species they chose and describing the habitat they created.

Mathematics Connection: Choose one endangered animal species in your area for the class to collect numerical data about. From the information you compiled, work as a class to describe this species in your area numerically (for example, how many individuals there were in the area before the species was endangered and how many now; how much of the animal's habitat has been destroyed).

Explanation

Science and Social Studies Classes and ELA Connection: Students will investigate engineering through an interactive read aloud. Remind students that they acted as engineers as they created their model habitats in the Habitat Helpers activity and that engineers are workers who design and build solutions to problems. Ask students to share what they know about engineers and what they wonder about engineers, recording students' ideas on a KWL chart.

Conduct an interactive read aloud of *Rosie Revere, Engineer* by Andrea Beaty. Before reading, remind students of the steps of the EDP and ask them to watch for these steps during the reading. After the read aloud, ask students to share what they learned about engineers, recording students' responses on the KWL chart. Have students create a STEM Research Notebook entry.

> #### STEM Research Notebook Entry #11
>
> Have students document what they learned about engineers and how they work in their STEM Research Notebooks, using both words and pictures.

Mathematics Connection: Ask students for their ideas about how engineers use mathematics. Create a class list of students' ideas.

Elaboration/Application of Knowledge

Science and Social Studies Classes and ELA Connection: Introduce the concept that animals and plants living in an area interact with each other and with other parts of their surroundings that are not living (for example air, water, rocks). Use as an example one animal that students identified in the Neighborhood Explorers activity (a bird, for example) and ask students for their ideas about how that animal interacts with other living things in the environment (birds use small twigs from trees for nests, eat worms). Then ask students how the animal interacts with nonliving things in the environment (birds fly through the air, drink water that collects in low areas of rocks). Introduce the term *interdependence* and the idea that animals and plants interact with each other and with nonliving things in their habitats in many ways, and that we call the area in which all these animals, plants, and nonliving things interact with each other an ecosystem.

Ask students for their ideas about the consequences for an ecosystem of an animal or plant becoming extinct, creating a class list of students' ideas. Conduct an interactive read aloud of *Who Eats What? Food Chains and Food Webs* by Patricia Lauber. After the read aloud, ask students to add to the list of consequences of animal or plant extinction, using what they learned from the read aloud. Next, choose an animal that students are familiar with and, as a class, create a simple linear diagram of that animals' food chain (for example, grass → cow → human or leaves → earthworm → robin). Have students complete a STEM Research Notebook entry in which they illustrate an example of a food chain for an animal that they are familiar with.

STEM Research Notebook Entry #12

Have students document their learning about food chains by drawing and labeling a food chain.

Next, connect students' understanding of their local habitat and endangered species with specific careers such as ecologist, biologist, park ranger, or naturalist. You may wish to invite a guest speaker to discuss endangered species and related careers with students.

Assess student learning by having students draw and label a picture of a local endangered animal species, and write one reason why this animal is endangered, one way this animal interacts with its surroundings, and one thing that humans can do to prevent this species from becoming extinct.

Mathematics Connection: Not applicable.

Evaluation/Assessment

Students may be assessed on the following performance tasks and other measures listed.

Performance Tasks

- Mighty Mobiles

- Habitat Helpers model and presentation

- Lesson assessment

Other Measures (using assessment rubric in Appendix B)

- Teacher observations

- STEM Research Notebook entries

- Participation in teams during investigations

INTERNET RESOURCES

EPA endangered species

- *www.epa.gov/endangered-species/learn-more-about-threatened-and-endangered-species*

List of endangered species by state

- *www.fws.gov/endangered/*

American bison

- *www.doi.gov/blog/15-facts-about-our-national-mammal-american-bison*

Engineering careers

- *www.engineergirl.org/33/TryOnACareer*

- *www.sciencekids.co.nz/sciencefacts/engineering/typesofengineeringjobs.html*

EDP

- *www.sciencebuddies.org/engineering-design-process/engineering-design-compare-scientific-method.shtml*

- *www.pbslearningmedia.org/resource/phy03.sci.engin.design.desprocess/what-is-the-design-process*

"Meet the American bison" video

- *www.youtube.com/watch?v=2iVO5mkIn1c*

Figure 4.1. Engineering Design Process

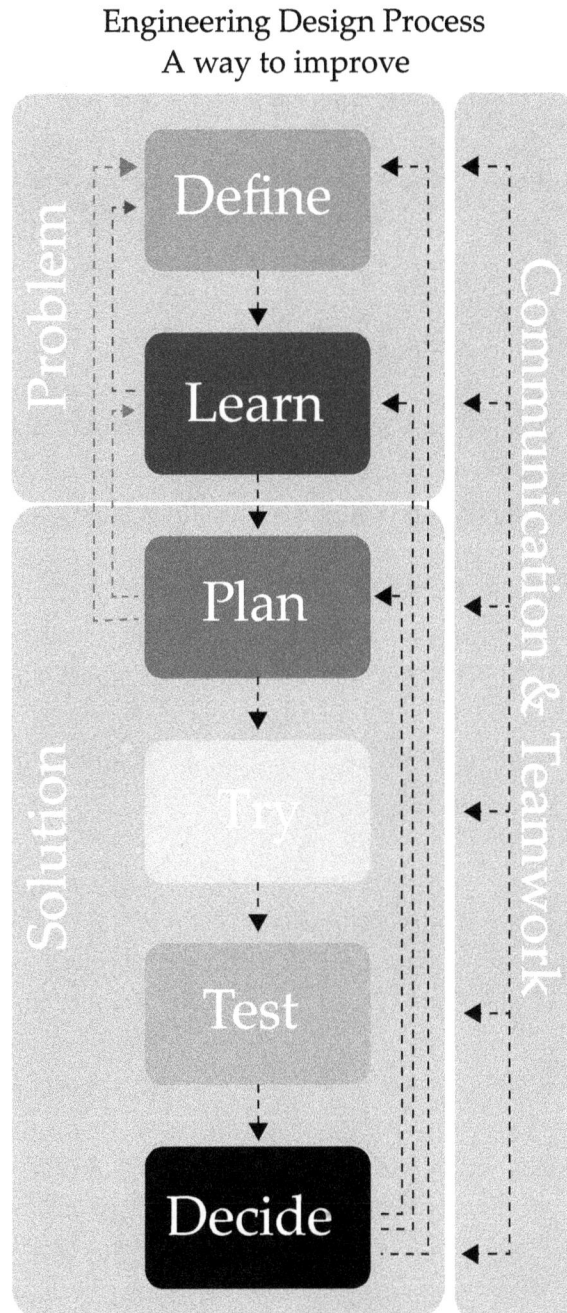

Engineering Design Process
A way to improve

©2015 PictureStem, Purdue University Research Foundation

Figure 4.2. American bison

Lesson Plan 3:
Endangered Species Throughout the World

In this lesson, students will explore various habitats and endangered species around the world. Student teams will choose one endangered species, either from another continent or from a North American region outside of their home region, and will use the EDP to design and build a model of this species, highlighting the animal's physical adaptations to its habitat.

ESSENTIAL QUESTIONS

- What is a physical adaptation?
- What continent is the endangered species you are studying from?
- What is the habitat like where your species live?
- What other types of animals live in the area where your endangered species live?
- What types of plants live in the continent you are studying?

ESTABLISHED GOALS AND OBJECTIVES

At the conclusion of this lesson, students will be able to do the following:

- Identify endangered species from parts of the world outside their home region
- Identify several reasons why species become endangered in other parts of the world
- Identify the continents and locate them on a map
- Identify and describe several animal physical adaptations
- Use the EDP to design and construct models to reflect how animals use their external parts to help them survive, grow, and meet their needs

TIME REQUIRED

7 days (approximately 30 minutes each; see Tables 3.8–3.9, pp. 36–37)

MATERIALS

Required Materials for Lesson 3

- STEM Research Notebooks
- Computer with internet access for viewing videos

- Books:
    - o *What do you do with a tail like this?* by Steve Jenkins
    - o Africa: *Amazing gorillas!* by Sarah L. Thomson
    - o Antarctica: *Big blue whale* by Nicola Davies
    - o Asia: *Save the orangutan* by Sarah Eason
    - o Australia: *Koalas* by Laura Marsh
    - o Europe: *Seals* by Wildlife in Danger
    - o North America: *The buffalo are back* by Jean Craighead George
    - o South America: *Endangered butterflies* by Bobbie Kalman

- Chart paper

- Markers

- U.S. map and world globe

- Yardsticks (1 for each pair of students)

Additional Materials for Amazing Adaptations Activity (for each team of 3–4 students)

- 2 paper plates

- 3 feet of yarn/string

- Access to paint and paint brushes

- 8 ounces of modeling clay

- 3 sheets of tissue paper, assorted colors

- 2 non-latex balloons

- Glue

- 3 sheets of construction paper, various colors

- 2 pairs of scissors

- 10 pony beads, various colors

SAFETY NOTES

1. Students should use caution when handling scissors, as the sharp points and blades can cut or puncture skin.

2. Have students wash hands with soap and water after the activity is completed.

CONTENT STANDARDS AND KEY VOCABULARY

Table 4.7 lists the content standards from the *NGSS*, *CCSS*, NAEYC, and the Framework for 21st Century Learning that this lesson addresses, and Table 4.8 (p. 80) presents the key vocabulary. Vocabulary terms are provided for both teacher and student use. Teachers may choose to introduce some or all of the terms to students.

Table 4.7. Content Standards Addressed in STEM Road Map Module Lesson 3

NEXT GENERATION SCIENCE STANDARDS

PERFORMANCE OBJECTIVES
- LS1–2. Read texts and use media to determine patterns in behavior of parents and offspring that help offspring survive.
- K-2 ETS1–2. Develop a simple sketch, drawing, or physical model to illustrate how the shape of an object helps it function as needed to solve a given problem.

DISCIPLINARY CORE IDEAS

LS1.A. Structure and Function
- All organisms have external parts. Different animals use their body parts in different ways to see, hear, grasp objects, protect themselves, move from place to place, and seek, find, and take in food, water and air. Plants also have different parts (roots, stems, leaves, flowers, fruits) that help them survive and grow.

LS1.B. Growth and Development of Organisms
- Adult plants and animals can have young. In many kinds of animals, parents and the offspring themselves engage in behaviors that help the offspring to survive.

LS1.D. Information Processing
- Animals have body parts that capture and convey different kinds of information needed for growth and survival. Animals respond to these inputs with behaviors that help them survive. Plants also respond to some external inputs.

CROSSCUTTING CONCEPTS

Patterns
- Patterns in the natural and human designed world can be observed, used to describe phenomena, and used as evidence.

Structure and Function
- The shape and stability of structures of natural and designed objects are related to their function(s).

Continued

Table 4.7. (*continued*)

SCIENCE AND ENGINEERING PRACTICES

Constructing Explanations and Designing Solutions
- Constructing explanations and designing solutions in K–2 builds on prior experiences and progresses to the use of evidence and ideas in constructing evidence-based accounts of natural phenomena and designing solutions.
- Use materials to design a device that solves a specific problem or a solution to a specific problem.

Obtaining, Evaluating, and Communicating Information
- Obtaining, evaluating, and communicating information in K–2 builds on prior experiences and uses observations and texts to communicate new information.
- Read grade-appropriate texts and use media to obtain scientific information to determine patterns in the natural world.

Developing and Using Models
- Modeling in K–2 builds on prior experiences and progresses to include using and developing models (i.e., diagram, drawing, physical replica, diorama, dramatization, storyboard) that represent concrete events or design solutions.
- Use a model to represent relationships in the natural world.

COMMON CORE STATE STANDARDS FOR MATHEMATICS

MATHEMATICAL PRACTICES
- MP1. Make sense of problems and persevere in solving them.
- MP2. Reason abstractly and quantitatively.
- MP3. Construct viable arguments and critique the reasoning of others.
- MP4. Model with mathematics.
- MP5. Use appropriate tools strategically.
- MP6. Attend to precision.
- MP7. Look for and make use of structure.
- MP8. Look for and express regularity in repeated reasoning.

MATHEMATICAL CONTENT
- NBT.B.3. Compare two two-digit numbers based on meanings of the tens and ones digits, recording the results of comparisons with the symbols >, =, and <.
- NBT.C.5. Given a two-digit number, mentally find 10 more or 10 less than the number, without having to count; explain the reasoning used.
- NBT.C.6. Subtract multiples of 10 in the range 10–90 from multiples of 10 in the range 10–90 (positive or zero differences), using concrete models or drawings and strategies based on place value, properties of operations, and/or the relationship between addition and subtraction; relate the strategy to a written method and explain the reasoning used.
- 1.MD.C.4. Organize, represent, and interpret data with up to three categories; ask and answer questions about the total number of data points, how many in each category, and how many more or less are in one category than in another.

- OA.A.1. Use addition and subtraction within 20 to solve word problems involving situations of adding to, taking from, putting together, taking apart, and comparing, with unknowns in all positions.
- OA.A.2. Solve word problems that call for addition of three whole numbers whose sum is less than or equal to 20, e.g., by using objects, drawings, and equations with a symbol for the unknown number to represent the problem.

COMMON CORE STATE STANDARDS FOR ENGLISH LANGUAGE ARTS

READING STANDARDS

- RI.1.1. Ask and answer questions about key details in a text.
- RI.1.2. Identify the main topic and retell key details of a text.
- RI.1.3. Describe the connection between two individuals, events, ideas, or pieces of information in a text.
- RI.1.7. Use the illustrations and details in a text to describe its key ideas.

WRITING STANDARDS

- W.1.2. Write informative/explanatory texts in which they name a topic, supply some facts about the topic, and provide some sense of closure.
- W.1.6. With guidance and support from adults, use a variety of digital tools to produce and publish writing, including in collaboration with peers.
- W.1.7. Participate in shared research and writing.
- W.1.8. With guidance and support from adults, recall information from experiences or gather information from provided sources to answer a question.

SPEAKING AND LISTENING STANDARDS

- SL.1.1. Participate in collaborative conversations with diverse partners aboutgrade 1 topics and textswith peers and adults in small and larger groups.
- SL.1.1.A. Follow agreed-upon rules for discussions.
- SL.1.1.B. Build on others' talk in conversations by responding to the comments of others through multiple exchanges.
- SL.1.1.C. Ask questions to clear up any confusion about the topics and texts under discussion.
- SL.1.3. Ask and answer questions about what a speaker says in order to gather additional information or clarify something that is not understood.
- SL.1.5. Add drawings or other visual displays to descriptions when appropriate to clarify ideas, thoughts, and feelings.

NATIONAL ASSOCIATION FOR THE EDUCATION OF YOUNG CHILDREN STANDARDS

- 2.E.1. Arrange firsthand, meaningful experiences that are intellectually and creatively stimulating, invite exploration and investigation, and engage children's active, sustained involvement by providing a rich variety of material, challenges, and ideas.
- 2.F.3. Extend the range of children's interests and the scope of their thought, present novel experiences and introduce stimulating ideas, problems, experiences, or hypotheses.

Continued

Table 4.7. (*continued*)

- 2.F.6. Enhance children's conceptual understanding through various strategies, including intensive interview and conversation, encourage children to reflect on and "revisit" their experiences.
- 2.G.2. Scaffolding takes on a variety of forms.
- 2.J.1. Incorporate a wide variety of experiences, materials and equipment, and teaching strategies to accommodate the range of children's individual differences in development, skills and abilities, prior experiences, needs, and interests.
- 3.A.1. Teachers consider what children should know, understand, and be able to do across the domains.

FRAMEWORK FOR 21ST CENTURY LEARNING
- Interdisciplinary Themes
- Learning and Innovation Skills
- Information, Media and Technology Skills
- Life and Career Skills

Table 4.8. Key Vocabulary in Lesson 3

Key Vocabulary	Definition
adaptation	features that animals and plants have that help them to live in their habitats
measurement	a number that shows the size or amount of something
unit	an amount that is used as a standard for measuring

TEACHER BACKGROUND INFORMATION
Endangered Species

The focus of this lesson will be on endangered species globally. As many as 40,000 species of plants and animals worldwide were threatened with extinction as of 2022 (see *https://worldanimalfoundation.org/wild-earth* for more information). The World Wildlife Fund (WWF) noted in its "Living Planet Report" that as of 2018 the size of animal populations globally had shrunk by 69% since 1970 (see *https://livingplanet.panda.org/en-US*). The WWF attributes the rapid increase in population depletion and species extinction to the rapid increase in human consumption and related effects such as deforestation, over-fishing, and pollution. For additional information about endangered species and conservation efforts globally, see the following websites:

- *www.worldwildlife.org/initiatives/wildlife-conservation*

- *www.iucn.org/theme/species*

- *www.animalplanet.com/show/extinct-or-alive-animal-planet*

Animal Adaptations

In this lesson, students will be introduced to the idea that animals have adaptations that make them well suited to live in a particular environment. Animals exhibit both behavioral and physical adaptations. Behavioral adaptations include behaviors such as migration and hibernation. Physical adaptations include features such as birds' beaks, mammals' fur, and monkeys' prehensile tails. Physical adaptations will be the focus of this lesson. Students should understand that these physical traits that are displayed by all or most of the individuals in a species in a specific area. The focus in this lesson should be on the ways that animals' bodies are good fits for the environmental conditions in which they live. The National Wildlife Federation's Ranger Rick program provides videos of animals in nature that you may wish to incorporate into this lesson (*www.rangerrick.org/videos*). For more information about adaptations, see *https://education.nationalgeographic.org/resource/adaptation-and-survival*.

COMMON MISCONCEPTIONS

Students will have various types of prior knowledge about the concepts introduced in this lesson. Table 4.9 outlines some common misconceptions students may have concerning these concepts. Because of the breadth of students' experiences, it is not possible to anticipate every misconception that students may bring as they approach this lesson. Incorrect or inaccurate prior understanding of concepts can influence student learning in the future, however, so it is important to be alert to misconceptions such as those presented in the table.

Table 4.9. Common Misconceptions About the Concepts in Lesson 3

Topic	Student Misconception	Explanation
Adaptations	Individuals can adapt to changes in their environment and pass on these adaptations to their offspring.	Adaptations emerge in populations over long periods of time; those individuals who have traits that are best suited to survival in their environments are most likely to survive and pass those traits along.
Endangered species	If an animal or plant species becomes extinct it will not affect humans.	We rely on plants and animals in our habitats for many things, including food production, medicines, and fresh air. When animals and plants become extinct, it changes the balance of our ecosystem.

Continued

Table 4.9. (*continued*)

Topic	Student Misconception	Explanation
	Pollution is the primary reason why animals become endangered.	Loss of habitat is actually the primary reason for species becoming endangered and extinct. The growing human population takes up increasing amounts of space, and this means that we change habitats in ways that could be harmful to other life as we adapt them to human life.

PREPARATION FOR LESSON 3

Review the Teacher Background Information provided, assemble the materials for the lesson, and preview the videos recommended in the Learning Plan Components section below.

In this lesson, each team of three to four students will choose an endangered species to investigate. This species should be located on another continent or in a North American region outside of the students' home region. You should compile a list of one endangered species per continent in advance from a website such as the World Wildlife Fund's species directory, found at *www.worldwildlife.org/species/directory?direct ion=desc&sort=extinction_status*. Have pictures available of each animal. You may wish to engage adult volunteers for additional support as students work in their teams to learn about an endangered species.

For the Introductory Activity/Engagement, choose one animal per continent and display those animals' pictures. Students will experience animals using webcams during this lesson, so you may wish to choose species that can be viewed via webcam. Examples of animal webcams include the following:

- *www.mangolinkcam.com*

- *https://nationalzoo.si.edu/webcams*

An option for this lesson is a zoo field trip. If you choose to include a zoo field trip, you may wish to ensure that students' species choices are accessible at the zoo you will visit.

If students have access to computers with internet access, review webcam sites and sites that provide grade-appropriate information about each of the animals and bookmark them for student research. If students will not have access to computers with internet during this lesson, you should print information about each of the endangered species and have books available in the classroom for students to use for their research. If you choose to take a zoo field trip, make preparations for the trip.

LEARNING PLAN COMPONENTS
Introductory Activity/Engagement

Connection to the Challenge: Begin each day of this lesson by directing students' attention to the module challenge, the Save the Species Challenge:

> There are hundreds of endangered species all over the world. If people don't do something, some of these species may become extinct. Your challenge is to develop an action plan about what people can do to help this species survive in its habitat.

Hold a brief class discussion of how students' learning in the previous days' lessons contributed to their ability to complete the challenge, and add students' responses to the "learned" section of the KWL chart.

Science and Social Studies Classes and ELA Connection: Review the seven continents with students using a globe or world map and make a list of the continents on the board. Next to this list, create a list of one species from each continent. Ask the class to match each species to its continent. Ask students what information they used to decide what continent the animals were from. Point out to students that animals' appearance can sometimes provide clues about where they live; for example, polar bears have thick coats of white fur that blend in with a snowy landscape, providing clues that these animals live where it is cold for much of the year. Remind students of the discussion about adaptations from Lesson 1. Have students name adaptations they see for each animal. As the class discusses each animal, move the picture to a separate piece of chart paper labeled with the continent where the animal lives, and list the physical adaptations students point out below the animal's picture.

Next, conduct an interactive read aloud of *What Do You Do with a Tail Like This?* by Steve Jenkins. After the read aloud, discuss what animals students learned about and what adaptations these animals had that helped them live in their environments, creating a class list.

Mathematics Connection: Not applicable.

Activity/Investigation

Science and Social Studies Classes and ELA Connection: Students will complete the Amazing Adaptations investigation.

Amazing Adaptations

Introduce the Amazing Adaptations investigation by telling students that they are going to research an endangered species and then create a model of the animal. Divide

the class into teams of three to four students each and assign each team one of the endangered species you displayed during the Introductory Activity/Engagement (note: if there are more than 28 students in a class you may need to assign an animal to more than one team).

Tell the students that they will use the EDP to conduct their research and make their models. Students will use craft materials to create their model of the endangered species they are studying. The models should include at least one physical adaptation (for example, a Tasmanian forester kangaroo from the Australian continent uses its long tail for balance and a prop for standing; students could make a clay figure that highlights the long tail of the kangaroo).

Review the steps of the EDP with students. Have teams conduct research about their endangered species using the internet or printed information and webcam observations. Guide students through STEM Research Notebook entries 13–16 to structure their work. Entry #13 represents the Define phase of the EDP, entry #14 the Learn phase, entry #15 the plan phase, and #16 the try, test, and decide phases. Each student should complete the STEM Research Notebook entries.

STEM Research Notebook Entry #13

Have students identify their endangered species and describe its habitat, using both words and pictures.

STEM Research Notebook Entry #14

Have students identify at least one physical adaptation that helps their species survive in its environment. Also have students record a reason why this species is endangered and list one thing that humans can do to help ensure that this species survives.

After students complete the learn phase, show students the set of materials they will have to create their models.

STEM Research Notebook Entry #15

Have students create drawings of their models, labeling the materials they would use to build this model. Students should include at least one physical adaptation of their animal in their drawing.

After each student has created a drawing, have teams work together to decide on one student's drawing, or to create a new drawing incorporating elements of multiple students' drawings.

STEM Research Notebook Entry #16

Have student teams build their models and ensure that the model incorporates at least one physical adaptation. Have students identify any improvements they can make to the model.

Have each team present its model and describe the endangered species' physical adaptations.

Mathematics Connection: Display students' models at the front of the classroom. Ask students if the models give information about the animals' sizes in real life. Emphasize that models provide important information about the thing that they represent, but that they may not include all information. Show the pictures from the Introductory Activity/Engagement activity and work as a class to order the animals from smallest to largest. Ask students for their ideas about how to make it easier to determine the sizes of each animal, and introduce comparisons to a standard object as a way of expressing size (for example, whether a gorilla, elephant, polar bear, giraffe are taller or shorter than the classroom doorway; then whether they are taller or shorter than a table in your classroom). Create a class list expressing the animals' sizes as comparison (for example, the gorilla is shorter than the doorway but taller than the table).

Explanation

Science and Social Studies Classes and ELA Connection: Have student teams each share their Amazing Adaptations models with the class. Students should provide the following information when prompted:

- Name of their animal

- Continent where the animal lives

- One adaptation the animal has

- One reason the animal species is endangered

- One thing people can do to help the species to survive

Students will learn about one additional endangered species that was not included in the Amazing Adaptations activity through an interactive read aloud. Choose one of the books from the Materials list (p. 76). Ask students what they know about this animal and what they wonder about it, recording students' responses on a KWL chart. Then conduct an interactive read aloud. After the read aloud, have students share what they learned, adding to the KWL chart and creating a STEM Research Notebook entry.

<u>**STEM Research Notebook Entry #17**</u>

Have students document what they learned about the endangered species in their STEM Research Notebooks, using both words and pictures.

Mathematics Connection: As a class, review the EPA Endangered Species page at *www.epa.gov/endangered-species/endangered-species-save-our-species-information.* This page includes pictures of endangered species with links to information about that species. Choose several animals and read the descriptions together, asking students to be alert to how the animals are described using numbers (for example, the indigo snake can grow as long as *8 feet* long). Create a class list of the animals and how each is described using numbers.

After you have reviewed several animals, review the list and ask students for their observations about how numbers are used. Introduce the idea that some of the numbers are counting numbers, telling how many of the species there are, and some of them are measurements. Ask students how they know that a number gives a measurement (i.e., it includes a unit, such as feet or miles). Have students brainstorm the various units they are familiar with, creating a class list.

Elaboration/Application of Knowledge

Science and Social Studies Classes and ELA Connection: Remind students that they are going to create an action plan for the endangered species they chose. As a class, brainstorm ideas about what sort of information they should include in their action plans, prompting students to consider what threats to animal habitats they have learned about.

Assess student learning by having students draw and label pictures of two endangered species that teams presented. Students should describe the animal and what continent it lives on using both words and pictures. Have students identify the seven continents on a map.

Mathematics Connection: Conduct an interactive read aloud of *Twelve Snails to One Lizard* by Susan Hightower. The book introduces measurement equivalents. Have students devise a classroom "unit of measurement" (for example, measuring by fingertip width, foot length, pencil length) and have multiple students measure the same distance using this unit of measurement. Hold a class discussion on the importance of standard measurements and introduce the relationship between inches, feet, and yards. Provide students with several objects and have them identify the best unit (for example, show a pencil, a piece of poster board, a table, and a carpet or other large object). Provide students with yardsticks and have them measure objects in the classroom, recording their measurements in the most appropriate units.

Evaluation/Assessment

Students may be assessed on the following performance tasks and other measures listed.

Performance Tasks

- Amazing Adaptations models

- Amazing Adaptations team presentations

- Lesson Assessment

Other Measures (using assessment rubric in Appendix B)

- Teacher observations

- STEM Research Notebook entries

- Participation in teams during investigations

INTERNET RESOURCES

Endangered species information

- *https://livingplanet.panda.org/en-US/*

- *www.worldwildlife.org/initiatives/wildlife-conservation*

- *www.iucn.org/theme/species*

- *www.animalplanet.com/show/extinct-or-alive-animal-planet*

National Wildlife Federation Ranger Rick videos

- *www.rangerrick.org/videos*

Animal adaptations

- *https://education.nationalgeographic.org/resource/adaptation-and-survival/*

World Wildlife Fund species directory

- *www.worldwildlife.org/species/directory?direction=desc&sort=extinction_status*

Animal webcams

- *www.mangolinkcam.com*

- *https://nationalzoo.si.edu/webcams*

EPA endangered species webpage

- *www.epa.gov/endangered-species/endangered-species-save-our-species-information*

Lesson Plan 4:
The Save the Species Challenge

In this lesson, students will focus on conservation efforts that can help ensure the survival of endangered species. The class will choose one local endangered species and identify ways that they can be involved in conservation efforts on a local level. Student teams will each create an action plan to ensure the survival of a selected endangered species on a global scale. This plan will include a description of the species' habitat and how humans and other factors may influence this habitat and contribute to the species' vitality. The action plan will focus on actions that humans can take to ensure this species' survival.

ESSENTIAL QUESTIONS

- What are conservation efforts?

- What can we do to help conserve a local endangered species?

- How can we help to conserve an endangered species globally?

ESTABLISHED GOALS AND OBJECTIVES

At the conclusion of this lesson, students will be able to do the following:

- Apply their learning about habitats and endangered animals to create an action plan that describes habitats and an endangered species' interaction with that habitat

- Apply their learning about habitats and endangered animals to identify actions that people can take to ensure the survival of endangered species

TIME REQUIRED

5 days (approximately 30 minutes each; see Table 3.10, p. 37)

MATERIALS

Required Materials for Lesson 4

- STEM Research Notebooks

- Computer with internet access for viewing videos

- Books used in previous lessons

- Save the Species Action Plan Templates

- Chart paper
- Markers
- U.S. map and globe

Additional Materials for Class Conservation Plan (1 for each team of 3–4 students)

- 11 × 13 piece of poster board
- 2 pairs of scissors
- 1 set of markers
- 2 glue sticks

SAFETY NOTES

1. Students should use caution when handling scissors, as the sharp points and blades can cut or puncture skin.

2. Have students wash hands with soap and water after the activity is completed.

CONTENT STANDARDS AND KEY VOCABULARY

Table 4.10 lists the content standards from the *NGSS*, *CCSS*, NAEYC, and the Framework for 21st Century Learning that this lesson addresses, and Table 4.11 presents the key vocabulary.

Vocabulary terms are provided for both teacher and student use. Teachers may choose to introduce some or all of the terms to students.

Table 4.10. Content Standards Addressed in STEM Road Map Module Lesson 4

NEXT GENERATION SCIENCE STANDARDS
PERFORMANCE OBJECTIVES • LS1–2. Read texts and use media to determine patterns in behavior of parents and offspring that help offspring survive. • K-2 ETS1–2. Develop a simple sketch, drawing, or physical model to illustrate how the shape of an object helps it function as needed to solve a given problem.

Continued

Table 4.10. (*continued*)

DISCIPLINARY CORE IDEAS

LS1.A. Structure and Function
- All organisms have external parts. Different animals use their body parts in different ways to see, hear, grasp objects, protect themselves, move from place to place, and seek, find, and take in food, water and air. Plants also have different parts (roots, stems, leaves, flowers, fruits) that help them survive and grow.

LS1.B. Growth and Development of Organisms
- Adult plants and animals can have young. In many kinds of animals, parents and the offspring themselves engage in behaviors that help the offspring to survive.

LS1.D. Information Processing
- Animals have body parts that capture and convey different kinds of information needed for growth and survival. Animals respond to these inputs with behaviors that help them survive. Plants also respond to some external inputs.

CROSSCUTTING CONCEPTS

Patterns
- Patterns in the natural and human designed world can be observed, used to describe phenomena, and used as evidence.

Structure and Function
- The shape and stability of structures of natural and designed objects are related to their function(s).

SCIENCE AND ENGINEERING PRACTICES

Constructing Explanations and Designing Solutions
- Constructing explanations and designing solutions in K–2 builds on prior experiences and progresses to the use of evidence and ideas in constructing evidence-based accounts of natural phenomena and designing solutions.
- Use materials to design a device that solves a specific problem or a solution to a specific problem.

Obtaining, Evaluating, and Communicating Information
- Obtaining, evaluating, and communicating information in K–2 builds on prior experiences and uses observations and texts to communicate new information.
- Read grade-appropriate texts and use media to obtain scientific information to determine patterns in the natural world.

Developing and Using Models
- Modeling in K–2 builds on prior experiences and progresses to include using and developing models (i.e., diagram, drawing, physical replica, diorama, dramatization, storyboard) that represent concrete events or design solutions.
- Use a model to represent relationships in the natural world.

COMMON CORE STATE STANDARDS FOR MATHEMATICS

MATHEMATICAL PRACTICES
- MP1. Make sense of problems and persevere in solving them.
- MP2. Reason abstractly and quantitatively.
- MP3. Construct viable arguments and critique the reasoning of others.
- MP4. Model with mathematics.
- MP5. Use appropriate tools strategically.
- MP6. Attend to precision.
- MP7. Look for and make use of structure.
- MP8. Look for and express regularity in repeated reasoning.

MATHEMATICAL CONTENT
- NBT.B.3. Compare two two-digit numbers based on meanings of the tens and ones digits, recording the results of comparisons with the symbols >, =, and <.
- NBT.C.5. Given a two-digit number, mentally find 10 more or 10 less than the number, without having to count; explain the reasoning used.
- NBT.C.6. Subtract multiples of 10 in the range 10–90 from multiples of 10 in the range 10–90 (positive or zero differences), using concrete models or drawings and strategies based on place value, properties of operations, and/or the relationship between addition and subtraction; relate the strategy to a written method and explain the reasoning used.
- 1.MD.C.4. Organize, represent, and interpret data with up to three categories; ask and answer questions about the total number of data points, how many in each category, and how many more or less are in one category than in another.
- OA.A.1. Use addition and subtraction within 20 to solve word problems involving situations of adding to, taking from, putting together, taking apart, and comparing, with unknowns in all positions.
- OA.A.2. Solve word problems that call for addition of three whole numbers whose sum is less than or equal to 20, e.g., by using objects, drawings, and equations with a symbol for the unknown number to represent the problem.

COMMON CORE STATE STANDARDS FOR ENGLISH LANGUAGE ARTS

READING STANDARDS
- RI.1.1. Ask and answer questions about key details in a text.
- RI.1.2. Identify the main topic and retell key details of a text.
- RI.1.3. Describe the connection between two individuals, events, ideas, or pieces of information in a text.
- RI.1.7. Use the illustrations and details in a text to describe its key ideas.

WRITING STANDARDS
- W.1.2. Write informative/explanatory texts in which they name a topic, supply some facts about the topic, and provide some sense of closure.
- W.1.6. With guidance and support from adults, use a variety of digital tools to produce and publish writing, including in collaboration with peers.

Continued

Table 4.10. (*continued*)

- W.1.7. Participate in shared research and writing.
- W.1.8. With guidance and support from adults, recall information from experiences or gather information from provided sources to answer a question.

SPEAKING AND LISTENING STANDARDS
- SL.1.1. Participate in collaborative conversations with diverse partners aboutgrade 1 topics and textswith peers and adults in small and larger groups.
- SL.1.1.A. Follow agreed-upon rules for discussions.
- SL.1.1.B. Build on others' talk in conversations by responding to the comments of others through multiple exchanges.
- SL.1.1.C. Ask questions to clear up any confusion about the topics and texts under discussion.
- SL.1.3. Ask and answer questions about what a speaker says in order to gather additional information or clarify something that is not understood.
- SL.1.5. Add drawings or other visual displays to descriptions when appropriate to clarify ideas, thoughts, and feelings.

NATIONAL ASSOCIATION FOR THE EDUCATION OF YOUNG CHILDREN STANDARDS
- 2.E.1. Arrange firsthand, meaningful experiences that are intellectually and creatively stimulating, invite exploration and investigation, and engage children's active, sustained involvement by providing a rich variety of material, challenges, and ideas.
- 2.F.3. Extend the range of children's interests and the scope of their thought, present novel experiences and introduce stimulating ideas, problems, experiences, or hypotheses.
- 2.F.6. Enhance children's conceptual understanding through various strategies, including intensive interview and conversation, encourage children to reflect on and "revisit" their experiences.
- 2.G.2. Scaffolding takes on a variety of forms.
- 2.J.1. Incorporate a wide variety of experiences, materials and equipment, and teaching strategies to accommodate the range of children's individual differences in development, skills and abilities, prior experiences, needs, and interests.
- 3.A.1. Teachers consider what children should know, understand, and be able to do across the domains.

FRAMEWORK FOR 21ST CENTURY LEARNING
- Interdisciplinary Themes
- Learning and Innovation Skills
- Information, Media and Technology Skills
- Life and Career Skills

Table 4.11. Key Vocabulary in Lesson 4

Key Vocabulary	Definition
action plan	a suggestion about steps or activities that can be taken to accomplish a goal

TEACHER BACKGROUND INFORMATION

This lesson focuses students' attention on actions humans can take to ensure the survival of endangered species. Students will consider conservation both from a local and a global perspective. You will provide a short list of local endangered species from which the class will choose one and, as a class, create a plan to support this species' survival. Student teams will then each choose an endangered species on a global scale to investigate. Students will conduct research to collect information similar to the information they collected for the Amazing Adaptations activity; however, in this lesson, they will focus on actions humans can take to support the survival of the endangered species.

Conservation

Conservation efforts can take a variety of forms. These efforts can include habitat protection such as reducing deforestation or creating nature preserves, public education efforts such as programs at zoos that educate visitors about threats to animals and encourage them to donate money or become involved, and passing laws such as the Endangered Species Act. For more information about various conservation efforts, see the following websites:

- *www.worldwildlife.org/initiatives/wildlife-conservation*

- *www.aza.org/conservation-education*

- *www.nwf.org/Educational-Resources/Wildlife-Guide/Understanding-Conservation/Endangered-Species*

COMMON MISCONCEPTIONS

Students will have various types of prior knowledge about the concepts introduced in this lesson. Table 4.12 outlines a common misconception students may have concerning these concepts. Because of the breadth of students' experiences, it is not possible to anticipate every misconception that students may bring as they approach this lesson. Incorrect or inaccurate prior understanding of concepts can influence student learning in the future, however, so it is important to be alert to misconceptions such as that presented in the table.

PREPARATION FOR LESSON 4

Review the Teacher Background Information provided, assemble the materials for the lesson, and duplicate copies of the Save the Species Action Plan Template for each student. In Lesson 3 (Elaboration/Application of Knowledge section), students brainstormed about information they would include in an action plan to conserve endangered species. A template for the Save the Species Action Plan activity pages for this

Table 4.12. Common Misconception About the Concepts in Lesson 4

Topic	Student Misconception	Explanation
Endangered species	There is nothing people can do in their everyday lives to help species survive.	Protecting our Earth's biodiversity requires that all people make conscious decisions in their daily lives involving their consumption and resource use and, in addition, people can take actions such as donating money to conservation causes and joining local conservation movements. By making mindful decisions, people can impact endangered species' survival.

activity is provided at the end of this lesson plan; however, you may wish to add pages based upon students' ideas from Lesson 3. Be prepared to read aloud the Save the Species Action Plan instructions and provide support for students as they create their plans.

Prepare a shortlist of local endangered species (see the U.S. Fish and Wildlife Service at *www.fws.gov/endangered/*). The class will choose one of these species and create a class action plan to support this species' survival. Prepare printed information about this animal for student teams to use for research. Student teams will research the following topics (assign one topic per team): habitat, physical adaptations, reasons why the species is endangered, other animals that live near the species, what efforts humans can take to conserve this species. Teams will each prepare a poster with the information, so you should provide a picture of the animal for each team to use. You may wish to engage adult volunteers for additional support as students work in their teams to create their action plans.

For students' work on their Save the Species Action Plan, select endangered animal species from around the world for which information and pictures are readily available. If students have access to computers with internet access, bookmark webpages with information or, alternatively, prepare information about and pictures of each animal and have books on hand for students to use for their research.

If you choose to invite guests to observe students' presentations (see Elaboration/Application of Knowledge section), extend invitations and make appropriate preparations. Provide these guests with an overview of the project and provide samples of developmentally appropriate questions and comments.

LEARNING PLAN COMPONENTS

Connection to the Challenge: Begin each day of this lesson by directing students' attention to the module challenge, the Save the Species Challenge:

There are hundreds of endangered species all over the world. If people don't do something, some of these species may become extinct. Your challenge is to develop an action plan about what people can do to help this species survive in its habitat.

Hold a brief class discussion of how students' learning in the previous days' lessons contributed to their ability to complete the challenge, and add students' responses to the "learned" section of the KWL chart.

Introductory Activity/Engagement

Science and Social Studies Classes and Mathematics and ELA Connections: Ask students if they have ever visited a local animal shelter, zoo, or aquarium. Allow students to share their experiences. Ask them to name some of the animals they saw. Tell students that zoos and aquariums play a major role in animal conservation. Ask students to offer their ideas of what conservation is and guide students to an understanding that conservation is restoring or protecting something in the environment. Ask students to offer their ideas about how zoos and aquariums help in conservation. Highlight animals that students saw that represent endangered species (for example, elephants, gorillas, red pandas, rhinoceros, zebras, desert tortoises).

Remind students that their teams will choose an endangered species and develop a Save the Species Action Plan focusing on ways humans can impact this species' survival. First, however, the class will choose a local endangered species and create a display about the animal that is similar to the information zoos provide about animals that teaches people about the animal and how they can work to conserve it.

Class Conservation Plan

Show the class the list of local endangered species you prepared. Have the class vote to select an endangered species. Next, have student teams each research components of the action plan. For example, have Teams 1 and 2 research habitat, Team 3 research adaptations to the environment, Team 4 research reasons why the species is endangered, Team 5 research other animals that live near the species, and Teams 6 and 7 research what efforts humans can take to conserve this species. Have each team prepare and present a small poster representing their findings and post these around the room as the class's action plan. Have the class offer additional ideas of ways that they can help in conservation efforts for this species and add these to the plan.

Activity/Investigation

Science and Social Studies Classes and Mathematics and ELA Connections: Students will address the module challenge, the Save the Species Challenge, by creating an action plan.

The Save the Species Action Plan

Student teams will select an endangered species globally (outside of the U.S. or in a North American region outside of students' home region). Students will conduct research to prepare their action plans (see the Save the Species Action Plan template attached at the end of this lesson plan).

Explanation

Science and Social Studies Classes and ELA Connections: Invite other students from the school to see the Class Conservation Plan and have student teams present their posters, or post the plan in a hallway or public area.

Mathematics Connection: Not applicable.

Elaboration/Application of Knowledge

Science and Social Studies Classes and Mathematics and ELA Connections: Have student teams present their action plans, describing the species' habitat, reasons for its endangered status, and actions humans can take to support the species' survival. Ensure that each member of the team presents a portion of the action plan using his or her written action plan as a resource.

After each group presents its plan, hold a class discussion about the strengths of each group's plan and ask students to share their ideas about possible improvements to the plans. Have the presenting groups respond with ways that they could adapt their plan based on the discussion.

Evaluation/Assessment

Students may be assessed on the following performance tasks and other measures listed.

Performance Tasks

- Class Conservation Plan team poster
- Save the Species Action Plan
- Team presentations of Save the Species Action Plan

Other Measures (using assessment rubric in Appendix B)

- Teacher observations
- STEM Research Notebook entries
- Participation in teams during investigations

INTERNET RESOURCES

Conservation information

- *www.worldwildlife.org/initiatives/wildlife-conservation*

- *www.aza.org/conservation-education*

- *www.nwf.org/Educational-Resources/Wildlife-Guide/Understanding-Conservation/ Endangered-Species*

Endangered species by region

- *www.fws.gov/endangered/*

Save the Species Action Plan

Name:

_____

_____

OUR ENDANGERED SPECIES

Identify your team's endangered animal species:

_____

_____

Draw a picture of the endangered animal:

HABITAT

Where does your endangered animal live?

_____

_____

Color in the continent where your endangered animal lives.

Draw a picture of the habitat where your endangered species lives.

Food

Write and draw two things your endangered animal species eats.

Food 1.

_____

_____

Food 2.

_____

_____

Shelter

Draw and label a picture of your endangered animal's home.

Threats to Survival

Name two reasons why your animal species is endangered.

1.

_____

_____

_____

_____

_____

_____

2.

_____

_____

_____

_____

_____

_____

People Take Action

Write and draw three things that people can do to help your endangered species.

Action 1.

_____

_____

_____

_____

_____

_____

People Take Action

Action 2.

_____

_____

_____

_____

_____

_____

People Take Action

Action 3.

_____

- -

_____

_____

- -

_____

_____

- -

_____

4

Suggested Books

Baines, R. 2007. *Arctic tale*. Washington, DC: National Geographic Society.

Butterfield, M. 1999. *This is a hot and steamy place. It rains a lot. Where am I?* Mankato, MN: Thameside Press.

Cherry, L. 1990. *The great kapok tree*. New York: Scholastic.

Daniels, C., and Neon Squid 2023. *Koala: A first field guide to the cuddly marsupial from Australia*. New York: Neon Squid Publishers.

Davies, N. 1997. *Big blue whale*. Cambridge, MA: Candlewick Press.

Donald, L. 2001. *Endangered animals*. New York: Children's Press.

Eason, S. 2009. *Save the panda*. New York: The Rosen Publishing Group.

Eason, S. 2009. *Save the polar bear*. New York: The Rosen Publishing Group.

Evans, S. 2018. *Animal homes*. Washington, DC: National Geographic Publishers.

Gibbons, G. 1994. *Nature's green umbrella*. New York: HarperCollins.

Grady, T. 2021. *Polar bear: Fascinating facts for kids*. Osprey, FL: Dylanna Publishing.

Guiberson, B. 2010. *Earth: Feeling the heat*. New York: Henry Holt and Company.

Hoare, G. 2010. *Endangered animals: Discover why some of the world's creatures are dying out*. London: DK Publishers.

Marsh, L. 2014. *Koalas*. Washington, DC: National Geographic Society.

Martin, L. 2001. *Wildlife in danger: Seals*. Vero Beach, FL: Rouke Publishing.

Martin, L. 2001. *Wildlife in danger: Rhinoceros*. Vero Beach, FL: Rouke Publishing.

O'Connor, S. 2022. *The whale who ate plastic*. Wicklow, Ireland: Madra Rua Publishing.

Pattison, D. 2012. *Desert baths*. Mt. Pleasant, SC: Sylvan Dell Publishing.

Royston, R. 2019. *Save the giant panda*. Oxford: Raintree Publishing.

Taylor-Butler, C., and C. Clinton. 2023. *Save the blue whales*. New York: Philomel Books.

Thomas, S. 2005. *Amazing sharks!* New York: HarperCollins Publishers.

Thomson, S. 2006. *Amazing dolphins!* New York: HarperCollins Publishers.

Yezerski, T. 2011. *Meadowlands: A wetlands survival story*. New York: Farrar Straus Giroux.

REFERENCES

Koehler, C., M. A. Bloom, and A. R. Milner. 2015. The STEM road map for grades K-2. In *STEM Road Map: A framework for integrated STEM education*, ed. C. C. Johnson, E. E. Peters-Burton, and T. J. Moore, 41–67. New York: Routledge.

TRANSFORMING LEARNING WITH HABITATS LOCAL AND FAR AWAY AND THE *STEM ROAD MAP CURRICULUM SERIES*

Carla C. Johnson

This chapter serves as a conclusion to the Habitats Local and Far Away integrated STEM curriculum module, but it is just the beginning of the transformation of your classroom that is possible through use of the *STEM Road Map Curriculum Series*. In this book, many key resources have been provided to make learning meaningful for your students through integration of science, technology, engineering, and mathematics, as well as social studies and English language arts, into powerful problem- and project-based instruction. First, the Habitats Local and Far Away curriculum is grounded in the latest theory of learning for students in grade one specifically. Second, as your students work through this module, they engage in using the engineering design process (EDP) and build prototypes like engineers and STEM professionals in the real world. Third, students acquire important knowledge and skills grounded in national academic standards in mathematics, English language arts, science, and 21st century skills that will enable their learning to be deeper, retained longer, and applied throughout, illustrating the critical connections within and across disciplines. Finally, authentic formative assessments, including strategies for differentiation and addressing misconceptions, are embedded within the curriculum activities.

The Habitats Local and Far Away curriculum in the Sustainable Systems STEM Road Map theme can be used in single-content classrooms (e.g., mathematics) where there is only one teacher or expanded to include multiple teachers and content areas across classrooms. Through the exploration of the Habitats Local and Far Away lesson plans, students engage in a real-world STEM problem on the first day of

DOI: 10.4324/9781003450184-7

instruction and gather necessary knowledge and skills along the way in the context of solving the problem.

The other topics in the *STEM Road Map Curriculum Series* are designed in a similar manner. NSTA Press and Routledge have published additional volumes in this series for this and other grade levels, and have plans to publish more.

For an up-to-date list of volumes in the series, please visit *www.routledge.com/ STEM-Road-Map-Curriculum-Series/book-series/SRM* (for titles co-published by Routledge and NSTA Press), or *www.nsta.org/book-series/stem-road-map-curriculum* (for titles published by NSTA Press).

If you are interested in professional development opportunities focused on the STEM Road Map specifically or integrated STEM or STEM programs and schools overall, contact the lead editor of this project, Dr. Carla C. Johnson, Professor of Science Education at NC State University. Someone from the team will be in touch to design a program that will meet your individual, school, or district needs.

APPENDIX A

STEM RESEARCH NOTEBOOK TEMPLATES

MY STEM RESEARCH NOTEBOOK

HABITATS LOCAL AND FAR AWAY

NAME:

STEM Research Notebook #1 (Lesson Plan 1)

NAME: _____ DATE: _____

Draw and label one endangered animal, and write one thing you know about this animal.

_____

_____

_____

_____

STEM Research Notebook #2 (Lesson Plan 1)

NAME: _____ DATE: _____

I learned. . .

_____

_____
_____

_____

STEM Research Notebook #3 (Lesson Plan 1)

NAME: _____ DATE: _____

VOCABULARY WORDS

Word	Definition	Picture

STEM Research Notebook #4 (Lesson Plan 1)

NAME: _____ DATE: _____

Describe how each of an animal's basic needs are met in its habitat.

Food	
Water	
Shelter	
Air	
Space	

                    NATIONAL SCIENCE TEACHING ASSOCIATION

STEM Research Notebook #5 (Lesson Plan 1)

NAME: _____ DATE: _____

Five Needs

List the basic needs of all living things	Describe what happens if this need is not met

STEM Research Notebook #6 (Lesson Plan 1)

NAME: _____ DATE: _____

Neighborhood Explorer

Animal 1	Picture of how basic need is met
	Food
	Water
	Air
	Shelter
	Space

NATIONAL SCIENCE TEACHING ASSOCIATION

STEM Research Notebook #6 (Lesson Plan 1), page 2

NAME: _____ DATE: _____

Neighborhood Explorer

Animal 2	Picture of how basic need is met
	Food
	Water
	Air
	Shelter
	Space

STEM Research Notebook #7 (Lesson Plan 1)

NAME: _____ DATE: _____

I learned that animal species become endangered because. . . .

_____

_____

_____

_____

NATIONAL SCIENCE TEACHING ASSOCIATION

STEM Research Notebook #8 (Lesson Plan 2)

NAME: _____ DATE: _____

I learned. . .

_____

_____

_____

_____

STEM Research Notebook #9 (Lesson Plan 2)

NAME: _____ **DATE:** _____

Mighty Mobiles

What local endangered species are you studying?

_____

_____

Draw a picture of your endangered species.

NATIONAL SCIENCE TEACHING ASSOCIATION

STEM Research Notebook #9, page 2 (Lesson Plan 2)

NAME: _____ DATE: _____

Where does this species live?

_____

_____

_____

_____

Color the state where we live and this endangered species lives.

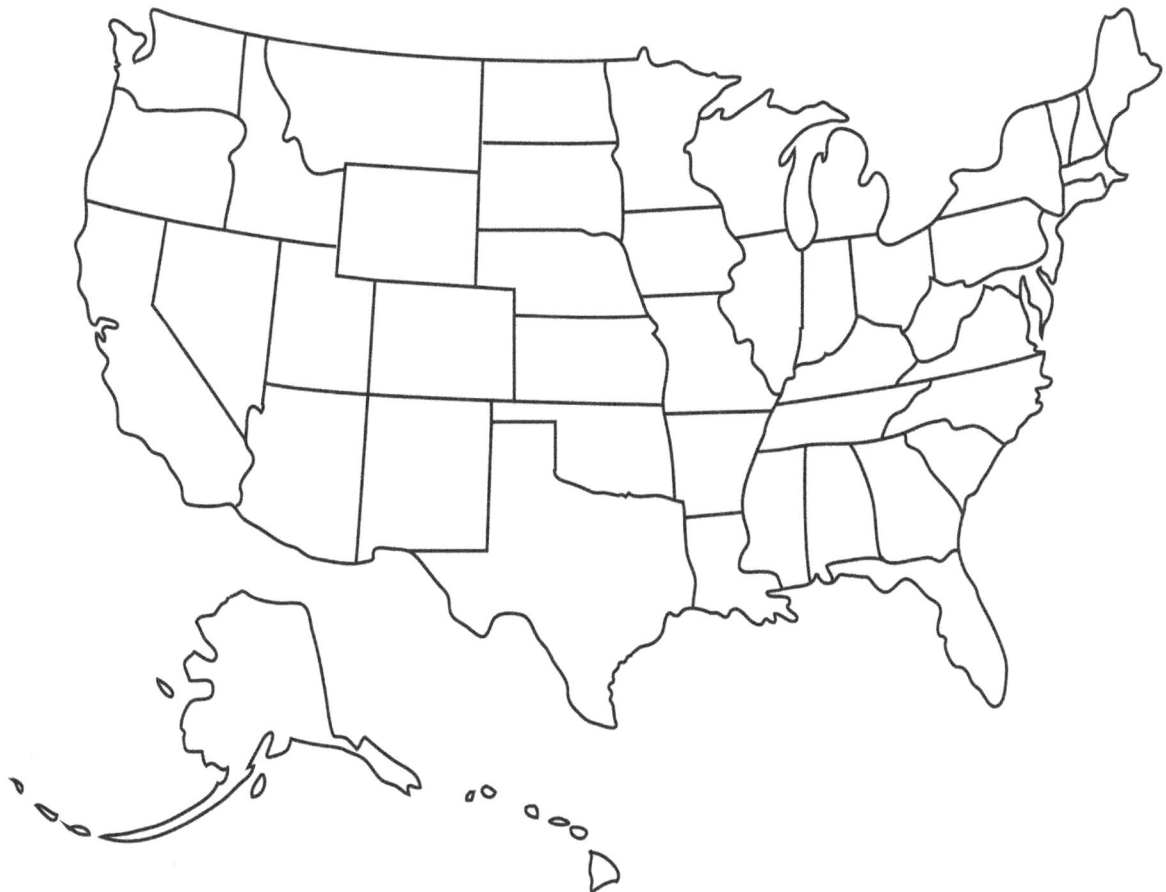

STEM Research Notebook #9, page 3 (Lesson Plan 2)

NAME: _____ DATE: _____

What is the endangered animal's habitat like?

_____

- -

_____

Draw a picture of the habitat where the endangered species lives.

STEM Research Notebook #9, page 4 (Lesson Plan 2)

NAME: _____ DATE: _____

List one reason this species is endangered.

_____

_____

_____

_____

_____

_____

_____

_____

_____

Name one thing people could do to help your endangered species.

_____

_____

_____

_____

_____

_____

_____

_____

_____

STEM Research Notebook #10 (Lesson Plan 2)

NAME: _____ DATE: _____

Habitat Helpers

DEFINE: What animal did your team choose?

_____

_____

LEARN: Think about what you need to know about the animal's habitat.

Where does this animal live in nature?

_____

_____

_____

_____

What are the animal's habitat needs (for example, does it need a dark place to live? Does it need to be high off the ground? Does it live in water or on land?)

_____

_____

_____

_____

_____

_____

STEM Research Notebook #10, page 2 (Lesson Plan 2)

NAME: _____ DATE: _____

PLAN: Draw the habitat you think your team should build. Remember that you have a set of materials to work with. Label the materials you will use on the drawing.

Your team must decide on ONE drawing to use to build your habitat. This might be one person's drawing or your team might choose to use parts of different people's drawings to make one final drawing. Sketch the final design your team decides on here.

STEM Research Notebook #10, page 3 (Lesson Plan 2)

NAME:_____ DATE:_____

TRY: Build your design based on your team's drawing

TEST/DECIDE: Look back at your animal's habitat needs (in the PLAN section). Will this habitat meet the animal's needs?

_____

_____

Can you do anything to improve your habitat design? If so, what can you do?

_____

_____

_____

_____

STEM Research Notebook #11 (Lesson Plan 2)

NAME: _____ DATE: _____

I learned. . .

_____

_____

_____

_____

STEM Research Notebook #12 (Lesson Plan 2)

NAME: _____ DATE: _____

Food Chain

Draw and label a food chain.

STEM Research Notebook #13 (Lesson Plan 3)

NAME: _____ DATE: _____

Amazing Adaptations – DEFINE

DEFINE:

- What endangered species are you are studying?

_____

_____

Draw a picture of your endangered species.

STEM Research Notebook #13, page 2 (Lesson Plan 3)

NAME: _____ **DATE:** _____

- Where does your endangered species live? (continent and country)

_____

_____

_____

_____

Color the continent where your endangered species lives.

STEM Research Notebook #13, page 3 (Lesson Plan 3)

NAME: _____ DATE: _____

- What is the habitat like where your endangered species lives?

_____

_____

Draw a picture of the habitat where your endangered species lives.

STEM Research Notebook #14 (Lesson Plan 3)

NAME: _____ DATE: _____

Amazing Adaptations – LEARN

LEARN – Adaptations:

- List one physical adaptation of your endangered species.

_____

_____

Draw a picture of your endangered species and label the physical adaptation.

NATIONAL SCIENCE TEACHING ASSOCIATION

STEM Research Notebook #14, page 3 (Lesson Plan 3)

NAME: _____ DATE: _____

LEARN – Why is this species endangered and how can people help?

- List one reason why this species is endangered.

_____

_____

_____

_____

_____

_____

_____

_____

_____

- List one thing people are doing to help this endangered species.

_____

_____

_____

_____

_____

_____

_____

_____

_____

STEM Research Notebook #15 (Lesson Plan 3)

NAME: _____ DATE: _____

Amazing Adaptations – PLAN

PLAN: What adaptations will you include in the model of your animal? Draw a picture of the model you would like to build and label at least one physical adaptation you will include. Label the materials you will use.

Your team must decide on ONE drawing to use to build your model. This might be one person's drawing or your team might choose to use parts of different people's drawings to make one final drawing. Sketch your team's final design on the back of this page.

STEM Research Notebook #16 (Lesson Plan 3)

NAME: _____ DATE: _____

Amazing Adaptations – TRY, TEST, and DECIDE

TRY: Build your design based on your team's drawing

TEST/DECIDE: Look back at your research on this animal's adaptations. Which one of these adaptations did you include in your model?

_____

_____

What can you do to improve your model?

_____

_____

_____

_____

Sketch your improvement here:

APPENDIX B

OBSERVATION, STEM RESEARCH NOTEBOOK, AND PARTICIPATION RUBRIC

OBSERVATION, STEM RESEARCH NOTEBOOK, AND PARTICIPATION RUBRIC

Name: _____   Date: _____

Categories (Components)	0 Missing or Unrelated	1 Beginning	2 Developing	3 Meets Expectation	4 Exceeds Expectation	TOTAL
Observation of Listening and Discussion Skills	Component is missing or unrelated.	Does not listen to others and shows little respect for alternative viewpoints.	Occasionally listens to others but often speaks out of turn.	Listens to others, only occasionally speaks out of turn, and generally accepts other points of view.	Listens carefully to others, waits for turn to speak, and respects alternative viewpoints.	
STEM Research Notebook	Component is missing or unrelated.	Indicates little understanding of the concepts being taught.	Recalls and is able to explain basic facts and concepts.	Demonstrates ability to apply concepts, using information in new situations.	Demonstrates a deep understanding of concepts by drawing relationships between ideas and using information to generate new ideas.	
Participation	Component is missing.	Does not volunteer. When responding to teacher prompts, comments are sometimes not relevant to the discussion.	Responds to teacher prompts during classroom discussions but does not volunteer.	Willingly participates in classroom discussions and offers relevant comments.	Contributes insightful comments and poses thoughtful questions.	
TOTAL						

COMMENTS

APPENDIX C

CONTENT STANDARDS ADDRESSED IN THIS MODULE

NEXT GENERATION SCIENCE STANDARDS

Table C1 (p. 144) lists the science and engineering practices, disciplinary core ideas, and crosscutting concepts this module addresses. The supported performance expectations are as follows:

- LS1–2. Read texts and use media to determine patterns in behavior of parents and offspring that help offspring survive.

- LS1–2. Read texts and use media to determine patterns in behavior of parents and offspring that help offspring survive.

- K-2 ETS1–2. Develop a simple sketch, drawing, or physical model to illustrate how the shape of an object helps it function as needed to solve a given problem.

Content Standards Addressed in STEM Road Map Module

Table C1. Next Generation Science Standards (NGSS)

SCIENCE AND ENGINEERING PRACTICES

CONSTRUCTING EXPLANATIONS AND DESIGNING SOLUTIONS

- Constructing explanations and designing solutions in K–2 builds on prior experiences and progresses to the use of evidence and ideas in constructing evidence-based accounts of natural phenomena and designing solutions.
- Use materials to design a device that solves a specific problem or a solution to a specific problem.

OBTAINING, EVALUATING, AND COMMUNICATING INFORMATION

- Obtaining, evaluating, and communicating information in K–2 builds on prior experiences and uses observations and texts to communicate new information.
- Read grade-appropriate texts and use media to obtain scientific information to determine patterns in the natural world.

DEVELOPING AND USING MODELS

- Modeling in K–2 builds on prior experiences and progresses to include using and developing models (i.e., diagram, drawing, physical replica, diorama, dramatization, storyboard) that represent concrete events or design solutions.
- Use a model to represent relationships in the natural world.

DISCIPLINARY CORE IDEAS

LS1.A. STRUCTURE AND FUNCTION

- All organisms have external parts. Different animals use their body parts in different ways to see, hear, grasp objects, protect themselves, move from place to place, and seek, find, and take in food, water and air. Plants also have different parts (roots, stems, leaves, flowers, fruits) that help them survive and grow.

LS1.B. GROWTH AND DEVELOPMENT OF ORGANISMS

- Adult plants and animals can have young. In many kinds of animals, parents and the offspring themselves engage in behaviors that help the offspring to survive.

LS1.D. INFORMATION PROCESSING

- Animals have body parts that capture and convey different kinds of information needed for growth and survival. Animals respond to these inputs with behaviors that help them survive. Plants also respond to some external inputs.

CROSSCUTTING CONCEPTS

PATTERNS

- Patterns in the natural world can be observed, used to describe phenomena, and used as evidence.

STRUCTURE AND FUNCTION

- The shape and stability of structures of natural and designed objects are related to their function(s).

CAUSE AND EFFECT

- Events have causes that generate observable patterns.

Source: NGSS Lead States. 2013. *Next Generation Science Standards: For states, by states.* Washington, DC: National Academies Press. *www.nextgenscience.org/next-generation-science-standards.*

Table C2. Common Core Mathematics and English Language Arts (ELA) Standards

Common Core State Standards for Mathematics	Common Core State Standards for English Language Arts
MATHEMATICAL PRACTICES MP1. Make sense of problems and persevere in solving them. MP2. Reason abstractly and quantitatively. MP3. Construct viable arguments and critique the reasoning of others. MP4. Model with mathematics. MP5. Use appropriate tools strategically. MP6. Attend to precision. MP7. Look for and make use of structure. MP8. Look for and express regularity in repeated reasoning. **MATHEMATICAL CONTENT** NBT.B.3. Compare two two-digit numbers based on meanings of the tens and ones digits, recording the results of comparisons with the symbols >, =, and <. NBT.C.5. Given a two-digit number, mentally find 10 more or 10 less than the number, without having to count; explain the reasoning used. NBT.C.6. Subtract multiples of 10 in the range 10–90 from multiples of 10 in the range 10–90 (positive or zero differences), using concrete models or drawings and strategies based on place value, properties of operations, and/or the relationship between addition and subtraction; relate the strategy to a written method and explain the reasoning used. 1.MD.C.4. Organize, represent, and interpret data with up to three categories; ask and answer questions about the total number of data points, how many in each category, and how many more or less are in one category than in another. OA.A.1. Use addition and subtraction within 20 to solve word problems involving situations of adding to, taking from, putting together, taking apart, and comparing, with unknowns in all positions. OA.A.2. Solve word problems that call for addition of three whole numbers whose sum is less than or equal to 20, e.g., by using objects, drawings, and equations with a symbol for the unknown number to represent the problem.	**READING STANDARDS** RI.1.1. Ask and answer questions about key details in a text. RI.1.2. Identify the main topic and retell key details of a text. RI.1.3. Describe the connection between two individuals, events, ideas, or pieces of information in a text. RI.1.7. Use the illustrations and details in a text to describe its key ideas. **WRITING STANDARDS** W.1.2. Write informative/explanatory texts in which they name a topic, supply some facts about the topic, and provide some sense of closure. W.1.6. With guidance and support from adults, use a variety of digital tools to produce and publish writing, including in collaboration with peers. W.1.7. Participate in shared research and writing. W.1.8. With guidance and support from adults, recall information from experiences or gather information from provided sources to answer a question. **SPEAKING AND LISTENING STANDARDS** SL.1.1. Participate in collaborative conversations with diverse partners aboutgrade 1 topics and textswith peers and adults in small and larger groups. SL.1.1.A. Follow agreed-upon rules for discussions. SL.1.1.B. Build on others' talk in conversations by responding to the comments of others through multiple exchanges. SL.1.1.C. Ask questions to clear up any confusion about the topics and texts under discussion. SL.1.3. Ask and answer questions about what a speaker says in order to gather additional information or clarify something that is not understood. SL.1.5. Add drawings or other visual displays to descriptions when appropriate to clarify ideas, thoughts, and feelings.

Source: National Governors Association Center for Best Practices and Council of Chief State School Officers (NGAC and CCSSO). 2010. *Common core state standards.* Washington, DC: NGAC and CCSSO.

Table C3. National Association for the Education of Young Children (NAEYC) Standards

NAEYC Curriculum Content Area for Cognitive Development: Science and Technology
2.E.1. Arrange firsthand, meaningful experiences that are intellectually and creatively stimulating, invite exploration and investigation, and engage children's active, sustained involvement by providing a rich variety of material, challenges, and ideas.
2.F.3. Extend the range of children's interests and the scope of their thought, present novel experiences and introduce stimulating ideas, problems, experiences, or hypotheses.
2.F.6. Enhance children's conceptual understanding through various strategies, including intensive interview and conversation, encourage children to reflect on and "revisit" their experiences.
2.G.2. Scaffolding takes on a variety of forms.
2.J.1. Incorporate a wide variety of experiences, materials and equipment, and teaching strategies to accommodate the range of children's individual differences in development, skills and abilities, prior experiences, needs, and interests.
3.A.1. Teachers consider what children should know, understand, and be able to do across the domains.

Source: National Association for the Education of Young Children (NAEYC). 2005. *NAEYC early childhood program standards and accreditation criteria: The mark of quality in early childhood education.* Washington, DC: NAEYC.

Table C4. 21st Century Skills from the Framework for 21st Century Learning

21st Century Skills	Learning Skills and Technology Tools	Teaching Strategies	Evidence of Success
Interdisciplinary Themes	Economic, Business, and Entrepreneurial Literacy Health Literacy Environmental Literacy	Provide students with the opportunity to investigate various habitats and endangered species, in the context of the business, economics and industry of everyday life (e.g., effects of human consumption, conservation efforts)	Students will incorporate their understanding of the impact of human activities on animals into their action plan to help preserve an endangered species.
Learning and Innovation Skills	Creativity and Innovation Critical Thinking and Problem Solving Communication and Collaboration	Facilitate creativity and innovation through the creation of habitat and animal models and an action plan that proposes ways people can help to preserve endangered animal species. Facilitate critical thinking and problem solving through use of the EDP and having students make observations about their surroundings. Support students in communication skills both within groups and in making presentation.	Students acquire and use deeper content knowledge as they work to complete models and their action plans. Students work collaboratively and communicate effectively in teams to complete and present a group project.

21st Century Skills	Learning Skills and Technology Tools	Teaching Strategies	Evidence of Success
Information, Media and Technology Skills	Information Literacy Media Literacy Information Communication and Technology Literacy	Engage students in guided practice and scaffolding strategies through the use of developmentally appropriate books, videos, and websites to advance their knowledge.	Students effectively use technology to acquire and use deeper content knowledge as they work to complete their action plans.
Life and Career Skills	Flexibility and Adaptability Initiative and Self-Direction Social and Cross-Cultural Skills Productivity and Accountability Leadership and Responsibility	Facilitate student collaborative teamwork to foster life and career skills.	Students collaborate to conduct research and work on group projects throughout the module.

Source: Partnership for 21st Century Learning. 2015. Framework for 21st Century Learning. *www.p21.org/our-work/p21-framework.*

Table C5. English Language Development Standards

English Language Development Standards: Grades pre-K-5 (WIDA 2020)
ELD Standard 1: Social and Instructional Language Multilingual learners narrate, inform, explain, and argue. ELD Standard 2: The Language of Language Arts Multilingual learners will interpret and construct informational texts in language arts with prompting and support. ELD Standard 3: The Language of Mathematics Multilingual learners will interpret and construct mathematical informational texts with prompting and support. ELD Standard 4: The Language of Science. Multilingual learners will interpret and construct scientific informational texts and explanations. ELD Standard 5: The Language of Social Studies Multilingual learners will interpret and construct informational texts in social studies.

Source: WIDA. 2020. *WIDA English language development standards framework, 2020 edition: Kindergarten–grade 12.* Board of Regents of the University of Wisconsin System. *https://wida.wisc.edu/sites/default/files/resource/WIDA-ELD-Standards-Framework-2020.pdf.*

INDEX

Page numbers in *italics* refer to figures, those in **bold** indicate tables.

NATIONAL SCIENCE TEACHING ASSOCIATION

For Product Safety Concerns and Information please contact our EU
representative GPSR@taylorandfrancis.com
Taylor & Francis Verlag GmbH, Kaufingerstraße 24, 80331 München, Germany

www.ingramcontent.com/pod-product-compliance
Lightning Source LLC
Chambersburg PA
CBHW061818210326
41599CB00034B/7034

9 7 8 1 0 3 2 5 8 4 6 7 6